STRUCTURED WALKTHROUGHS

STRUCTURED WALKTHROUGHS

Fourth Edition

Edward Yourdon

YOURDON Press
Prentice Hall
Englewood Cliffs, New Jersey 07632

Yourdon, Edward.
    Structured walkthroughs / Edward Yourdon. -- 4th ed.
        p.   cm. -- (Yourdon Press computing series)
    Bibliography: p.
    Includes index.
    ISBN 0-13-855289-4
    1. Structured walkthrough (Computer science)   I. Title.
    II. Series.
    QA76.9.D43Y68 1989
    005.1--dc19   88-22927
                                                              CIP

Editorial/production supervision: Sophie Papanikolaou
Cover design: Lundgren Graphics
Manufacturing buyer: Mary Ann Gloriande

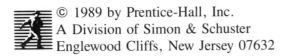 © 1989 by Prentice-Hall, Inc.
A Division of Simon & Schuster
Englewood Cliffs, New Jersey 07632

The publisher offers discounts on this book when ordered in bulk quantities. For
more information, write:

> Special Sales/College Marketing
> Prentice-Hall
> College Technical and Reference Division
> Englewood Cliffs, NJ 07632

Printed in the United States of America

10   9   8   7   6   5   4   3   2

ISBN   0-13-855289-4

Prentice-Hall International (UK) Limited, *London*
Prentice-Hall of Australia Pty. Limited, *Sydney*
Prentice-Hall Canada Inc., *Toronto*
Prentice-Hall Hispanoamericana, S.A., *Mexico*
Prentice-Hall of India Private Limited, *New Delhi*
Prentice-Hall of Japan, Inc., *Tokyo*
Simon & Schuster Asia Pte. Ltd., *Singapore*
Editora Prentice-Hall do Brasil, Ltda., *Rio de Janeiro*

CONTENTS

PREFACE

That was then, this is now. Technology changes, people don't. In 1977, when I wrote the first edition of this book, walkthroughs were regarded as an anathema by most programmers; and those few who did practice walkthroughs were typically concerned with removal of syntax errors in their code, for they were faced with overnight, or even week-long, delays in compiling their programs. Today, prototyping tools, end user development, JAD sessions, CASE tools, and PCs with mind-boggling power have completely changed the way systems are developed, and thus the way walkthroughs are carried out.

Perhaps the most important change is the widespread recognition that quality[1] is crucial in information systems, and that it doesn't appear all by itself. Possibly because of this recognition, I now find that approximately 75 percent of the MIS organizations in the United States use some form of walkthroughs at one or more stages in the development of their systems. And more is being written on walkthroughs and related technological issues: The bibliography in this new edition contains some 90 entries, more than double the number in the previous edition.

The widespread use of structured systems development techniques is another drastic change since this book was first written. Surveys in *Datamation* and other industry journals

---

[1] A high-quality information system is one that (a) carries out the purpose intended by the user, (b) can be built without schedule delays or cost overruns, (c) has no errors, and (d) can be maintained and modified to meet changing conditions at a minimal cost.

indicate that structured programming and structured design are used in over 75 percent of American MIS organizations; structured analysis and information modeling are also widely used, though not as widely as structured programming and structured design. Walkthroughs are a natural complement to the structured techniques. Many organizations find that the first productivity technique that should be introduced is the walkthrough, because it helps ensure that other techniques, such as structured analysis, are introduced and used by the staff in a consistent, uniform fashion.

Of course, the entire hardware/software environment has also changed drastically. Awesomely powerful personal computers are everywhere (including, thank goodness, on the desks of systems analysts and programmers!); local area networks and other communication protocols are finally linking all of these PCs into the mainframe environment; fourth generation languages (4GLs) and application prototyping tools are making it possible to develop working systems more quickly than ever before. The existence of these tools has led some organizations to use "playthroughs" instead of walkthroughs; a playthrough is a live simulation of some portion of a system, in which the various users play the roles that they intend to play in the "real" system.

One important consequence of powerful systems development tools is the possibility of smaller project teams. Instead of a Mongolian horde of 75 programmers and systems analysts toiling away for two years to produce 200,000 lines of COBOL, it is now possible for a small group of three to five people to produce the same system with 20,000 lines of code in a fourth generation language like FOCUS or RAMIS (or dBASE-III, or HyperTalk, or Mapper) in a period of six months. A group of 75 people has enormous difficulty managing all the communication necessary to make a project work, and it rarely functions as a socially cohesive team. A small group of three to five people, though, does have a definite opportunity to work closely together, perhaps in the same office, and thus become a programming team in the sense described in this book. So we may begin to see more true programming teams in the future, a trend which I view as a positive side-benefit of the fourth generation language movement.

Systems development teams have changed in another way: because of increased computer literacy and the widespread availability of PCs and 4GLs, teams are much more likely to include end users, either as working participants in the team or as supervisors of the team, or as system developers in their own right. Indeed, many project teams now work entirely within the user organization rather than hiding in a corner of the MIS empire. As a result, the walkthrough concept is even more important than before because it helps ensure better communication between the people who build the systems and the people who pay for and ultimately use the systems.

It is universally agreed today that effective communication between the user and the systems analyst is essential to the successful development of large, complex systems. To reflect this emphasis, I have changed the wording of some material in the book to highlight the work of systems analysts and to reduce the emphasis that previous editions placed on programmers and code walkthroughs.

The increased role of users is evident in other ways. The widespread availability of prototyping tools means that many walkthroughs involve the user/developer review of a prototype *product*, not the review of an abstract model (i.e., a picture or diagram) on a piece of paper. And the recent growth of expert systems means that many walkthroughs involve a review of the "knowledge acquisition" process that many users find difficult to articulate. I will comment on these special types of walkthroughs later in this book.

There is one other technological development that has affected the way that walkthroughs are used in some MIS organizations: the gradual proliferation of personal computers, terminals, office automation support, and on-line development tools. The newest of these, as we enter the 1990s, are CASE (**C**omputer **A**ssisted **S**oftware **E**ngineering) tools that allow the systems analyst to develop dataflow diagrams, entity-relationship diagrams, and other pictorial models of a system, and then automatically check the system models for syntactic completeness and consistency. With tools like these, much of the error-detection work that was done manually in a walkthrough can now be done quickly and efficiently by a computer.

However, this technology does not eliminate the need for walkthroughs! Instead, it allows the walkthrough participants to focus their attention on the "non-trivial" errors of style, maintainability, and so on. In fact, walkthroughs should now be much more interesting and productive than before, since the drudgery of looking for many types of errors (such as balancing errors in a dataflow diagram) can be eliminated.

CASE tools are just one example of the use of technology to improve the walkthrough process itself; indeed, the CASE tools of 1988 (when this book was written) are dramatically more powerful than the CASE tools of 1985, and they will undoubtedly continue to become more powerful throughout the early 1990s. In addition, increasingly sophisticated electronic mail, conferencing facilities, and electronic "post-it" comments attached directly to documents and diagrams will facilitate asynchronous walkthroughs among many geographically separated participants. And I believe that artificial intelligence and expert system technology will eventually provide system developers with an "expert reviewer" which will augment the review process carried out by humans. These and other issues are included in a new chapter on the future of walkthroughs as I think they will be carried out in the early and mid-1990s.

But ultimately humans are still the key ingredient: walkthroughs remain, just as they were ten years ago, semi-formal reviews among human peers who are attempting to improve the quality of their products. With a renewed interest in the quality of information systems around the world, walkthroughs should become a universally practiced technique in the industry.

As always, it is a pleasure to acknowledge the assistance I received during the preparation of this new edition. As Managing Editor for Yourdon Press, Ed Moura supervised the entire project from beginning to end. Bob Spurgeon, formerly Vice President of YOURDON inc.'s Federal Systems Division and now a globe-trotting independent consultant, covered my initial manuscript with his usual blunt (but very useful) comments and dropped it off enroute to a United Nations assignment in Kuwait. Mark Wallace, a Los Angeles based consultant, added his own corrections and suggested some valuable additions to the bibliography. Nan Matzke, a veteran systems analyst and practitioner of structured walkthroughs, annotated the manuscript with dozens of war stories and real-world experiences, which I found tremendously helpful. Roland Racko, a true Renaissance man in the systems development field, chided me for loose, sloppy thinking and forced me to tighten the material as best I could. Chet Delaney, of Chase Manhattan Bank, constantly reminded me that not all large MIS organizations are as screwed up as I portray them in this book. Finally, my wife Toni took the resulting revised manuscript and turned it into something resembling real English; without her, there would have been no book.

In the end, though, this book is *not* a team product in the sense discussed in Chapter 3. The reviews, comments, and suggestions created in the pseudo-walkthroughs of the manuscript have improved it enormously, but any remaining errors are mine alone.

A final note: As a child of the 1950s, I grew up unconsciously using "he" to refer to a person who could be either male or female; I envy the Germans, whose language includes the neuter pronoun "es" in addition to "er" and "sie." I find the repeated use of "he or she" and "s/he" to be clumsy and belabored, and thus continue to use "he." However, please keep in mind throughout the book that the word "he" is never meant to imply the exclusive purview of males, and that men and women can (and should) carry out equal roles in walkthroughs and all other aspects of the computer profession, as in life.

Edward Yourdon
New York City

STRUCTURED
WALKTHROUGHS

1 INTRODUCTION

Life is short, the art long, opportunity fleeting, experience treacherous, judgment difficult.

Hippocrates, *Aphorisms*, Section I

We are a nation in turmoil. During the past few years, educators, legislators, and a variety of experts have advised us that our highways are crumbling, our air is poisoned, and our government is paralyzed by its own bureaucracy. Our educational system is a shambles; our judicial system seems more successful at encouraging crime rather than preventing it. Industry is stagnating; worker productivity is actually declining in some industries; the quality of our country's products is inferior to that of the products of many other countries; and our American system of management is attacked almost universally as being out of step with the times.

To this long list of problems we must add a new one. It has to do with computer software and information systems. In addition to creating some difficulties of its own— difficulties that are large enough and serious enough that they deserve national attention— software compounds the problems listed above because computer systems now represent a critical component of almost every product we build and every service we provide.

A provocative book by Stephen McClellan [McClellan, 1984] estimates that in 1985, the information processing industry represented approximately 8 percent of the Gross

National Product of the United States. The industry is now twice as large as the steel industry and equal in size to the Big Three automobile manufacturers in the U.S. Only the oil industry is larger, and McClellan estimates that computers will overtake oil by the early 1990s.

The significant point about this development— which even people employed in the computer industry do not yet realize— is that the computer industry is already dominated by software. An average organization spending money on computers will find that about 50 percent of its money is spent on hardware; the other 50 percent is spent on software and on the people who produce and maintain that software. *By the early 1990s, the average organization will spend 80 percent of its computer budget on software and only 20 percent on hardware.* This ratio is already beginning to have a dramatic impact on the computer industry; indeed, McClellan estimates that by 1992 IBM's non-hardware revenues will exceed its hardware revenues.

These circumstances lead us to a central problem— and ultimately to the theme of this book: As a nation we do an absolutely dismal job at the business of producing software for our complex information systems. Our software is produced with unbelievably low productivity and with almost none of the predictability (for example, predictable schedules, budgets, and resource requirements) that we have come to expect of other engineering disciplines, where estimates are expected to be accurate to within ± 5 percent. Even worse, when we finish developing software for a complex information system, it usually doesn't work.

In addition, the cost of maintaining, enhancing, and improving old software is staggering: A survey in [Lientz and Swanson, 1980] estimates that most organizations spend approximately 50 percent of their software budget maintaining old systems, and many of the larger organizationsregularly spend 65-75 percent of their budget on maintenance.[1]

---

1 An even more dismal thought: Even if we could learn to develop software effectively by tomorrow morning, the problem wouldn't go away for at least another ten years. Estimates of the installed base of application programs (payroll, accounting, inventory,

These problems are not new. The computer industry and the people who program and design systems have known these statistics for a decade or more. But for many years, all of this was regarded as a technical issue, something to be discussed only among programmers and systems analysts. Finally, people in senior management positions in government and industry are beginning to realize that we are faced with a *national* problem. After all, the computer industry (which is now dominated by software) is one of the largest industries in the world, and is certainly one of the most rapidly growing industries; hence, it is a target for intense competition at the international level. Thus, unless we can learn to develop high-quality software at a competitive price, we run the serious risk of losing yet another industry to our neighbors in Europe or Asia or South America.

For more and more computer systems these days, the most important measure of quality is *correctness*. It is embarrassing, expensive, and often disastrous to allow a large program or system to be put into operation when it still contains defects. To some extent this always has been true, but today we find ourselves building increasingly more of the so-called "critical" systems, in which a single failure can lead to the loss of millions of dollars or hundreds of lives. Even for "non-critical" systems, software quality is becoming a major concern since (a) embedded computers are becoming a more common and critical component of products ranging from automobiles to dishwashers; and (b) software is almost always the largest, most expensive, and most troublesome component of those embedded systems. The Japanese, among others, are prepared to take advantage of this quality issue. Just as they captured the automobile market, the camera market, and a number of other consumer markets by combining high quality with a competitive price, so they now see an opportunity to do the same with computer software.

(1 cont.) etc.) in the United States range from $300 billion to $1 *trillion*; the U.S. Defense Department alone estimates that it has between 200 million and 300 million lines of installed code (we should be even more terrified by the fact that they don't know!). Worldwide, there are approximately 100 billion lines of installed code, of which approximately 70 billion are currently written in COBOL.

1.1   About This Book

This book is about a simple, practical, well-accepted technique for drastically improving the quality (in terms of economical development, reduced defects, and improved maintainability) of computer software: *walkthroughs*. It will tell you when to have them, how to conduct them, and how to avoid a variety of psychological and managerial problems that often accompany them.

But before we begin, there are several basic questions that need to be answered. What *is* a walkthrough? What do people hope to accomplish with a walkthrough? What can you hope to accomplish with this book? What assumptions does this book make about you?

This chapter answers these questions. If you already are familiar with the basic concept of walkthroughs, you can skip ahead to Chapter 2.

1.2   What Is a Walkthrough?

There is nothing mysterious about the basic concept of a walkthrough: It is simply a peer group review of any product. Throughout this book, we will assume that the product has something to do with a computer program or system; thus, I will be referring to walkthroughs of program listings, structure charts, dataflow diagrams, entity-relationship diagrams, and other models that are associated with the development of information systems. Alternatively, a walkthrough might be concerned with operational prototypes of a system in order to review the functionality, performance, or user interface.

As we will see in Chapter 2, walkthroughs can take place at various times in the development of a system. Also, a walkthrough can have a range of formats and can involve different groups of people. Despite the variation, the underlying activity remains the same: A group of peers— people at roughly the same level in the organization— meet to review and discuss a product. Because walkthroughs often are used in the development of complex systems within large, centralized MIS organizations, much of the discussion and many of the examples in this book will reflect this use. However, walkthroughs also can be used on small projects involving two or three programmer/analysts. And they can take place between system developers and end users, or among a group of end users who are building their own system.

In your organization and in articles you may have seen in the computing literature, walkthroughs may be referred to as code reviews, design reviews, or inspections. They also are called stompthroughs or walkovers because of some of the problems I will discuss in Parts III and IV of this book. In most cases, these terms are synonyms for walkthrough; in the few cases where different meanings are intended, I will give careful definitions.

1.3   Why Do People Practice Walkthroughs?

To a typical programmer or systems analyst, the notion of spending an hour reading through someone else's program listing or dataflow diagram makes no sense. Moreover, the thought of letting someone else look at his work strikes him as a waste of time, if not an invasion of privacy. This is even more true today in the world of microcomputers, where the industry extols the feats of lone "cowboy" programmers who write dazzling new programs— *all by themselves*— on the IBM PC or Macintosh computer.

Indeed, why would any rational programmer want someone else to look at his program listing? What is the point of a structured walkthrough? The answer is very simple: *Walkthroughs are an effective way to improve the quality of the source code of a computer program and the documents that describe the design, architecture, performance, user interface, or functional requirements of a system.*

Other, complementary approaches to software quality exist. There are software engineering techniques, such as structured analysis, structured design, and structured programming; testing techniques; automated tools for checking the consistency and correctness of computer programs; and techniques for developing rigorous mathematical proofs of correctness of computer programs. All these methods should be used whenever possible. Any approach that concentrates on introducing quality *before* the fact certainly is compatible with the idea of walkthroughs, and is preferable to many of the techniques for removing defects *after* the fact. For a thorough discussion of testing techniques and other approaches to software quality, [Myers, 1976], [Myers, 1979], [Dunn and Ullman, 1982], and [Dunn, 1984] are highly recommended.

Indeed, many after-the-fact techniques—the whole range of static testing and dynamic testing techniques used by MIS organizations, for example—now are widely recognized as failures. There is no such thing as exhaustive testing of a computer program (a fact that is not yet known to a new generation of end users building expert systems on their own personal computers!). And frequently tests that are performed merely reflect the same logical errors as the computer program itself, because they reflect the logical misunderstandings of the programmer. Similarly, the notion of generating rigorous mathematical proofs of correctness of "real-world" programs is appealing and is performed successfully on some medium-sized programs. However, the level of mathematical expertise, not to mention the cost (often approaching half a million dollars for a correctness proof of a few thousand lines of code), puts it far beyond the reach of the average organization. Even a successful proof of correctness has its limitations: All it can do is demonstrate that the code conforms to the written specifications, which may or may not match the user's real requirements for the system.

In contrast, walkthroughs have been found highly successful in helping to produce reliable, bug-free programs. Programming groups using walkthroughs report that they have been able to reduce the number of errors in production programs by as much as a factor of ten. In a typical organization, it is usual to find an average of three to five bugs in every hundred lines of code during the five- to-ten-year lifetime of a normal system. But in systems developed with modern software engineering methods, combined with diligent walkthroughs and other testing techniques, it is not uncommon to see as few as three to five bugs per *ten thousand* lines of code.[2]

Of course, coding errors are only part of the picture— usually only a small part.   The difficulty of maintaining computer systems also comes from design flaws, and even worse, from errors in analysis or requirements definition.   In some ways, such errors are far worse than bugs in the code: We can end up with a brilliant solution to the wrong problem. Recent studies (see, for example, [Martin, 1984]) indicate that, on average, 50 percent of the errors that are eventually detected in an operational system can be traced back to errors or misunderstandings in the systems analysis phase of the project.   More importantly, 75 percent *of the cost of error removal* in an operational system can be traced back to those same errors in the systems analysis phase.   Obviously, walkthroughs can be extremely helpful in this area, too.

Not only does the walkthrough approach find more errors than classical development techniques, but it also finds the errors more quickly and more economically.   I will elaborate upon this point in subsequent chapters, but the reason is quite simple: The author of any product, especially the author of a computer program, has mental blocks which prevent him from seeing errors in his product as quickly as a group of peers.   In addition, the walkthrough approach can often eliminate delays caused by slow turnaround for compilations, test shots, and so forth.   This is less of a problem in today's on-line, workstation-oriented systems development environment, though a depressingly large number of organizations still work in a batch environment.

As a result, organizations find that walkthroughs increase the productivity of their development staff.   This increase is not surprising, considering that roughly 50

---

2 Lest you think that this is an appropriate level of quality, let me point out that many researchers estimate that the U.S. Star Wars system, if ever built, may require as much as 100 million lines of code. This means that our "state of the art" level of quality would leave us with "only" 30,000-50,000 bugs, any one of which could obliterate life on this planet.

percent of the time spent on most development projects is spent on testing. In addition, diligent walkthroughs can help drastically reduce the effort required to maintain the system. After all, if there are no errors in the system, much less effort is required to keep the system running! Since maintenance occupies approximately 50-75 percent of the overall MIS budget in most organizations (see [Lientz and Swanson, 1980] for statistics about this) a reduction in the maintenance effort indirectly, but significantly, improves the MIS organization's ability to produce new systems.

In addition to detecting errors, walkthroughs help improve the overall quality of computer programs and systems. In particular, walkthroughs can help spot gross design or implementation inefficiencies. They can also help spot design strategies or coding techniques that would seriously hamper the system's maintainability. To the author of a program, it may not have been obvious that his method of accessing elements in a three-dimensional array would cause severe thrashing in a virtual memory environment; someone participating in a walkthrough will spot it. It may not have occurred to the author that his choice of data names is cryptic, or even totally meaningless to others; a walkthrough will bring that problem out. Similarly, the systems analyst who produces a dataflow diagram supposedly representing the essential requirements of a user's system may not realize that he has made subtle (and disastrous) assumptions about the kind of hardware/software technology that will be used to implement the system; once again, a walkthrough attended by other experienced and objective systems analysts can help spot this oversight.

Another way of ensuring the quality of a program is by establishing *standards* for the analysis, design, coding, testing, and documentation of programs. While standards are important, situations can occur when someone has to interpret them properly so that the implementer obeys the spirit and not just the letter of the standards. Unfortunately, the individual doing the work is not always in a good position to make this judgment.

For example, if the standards indicate that complex nested IF statements should be avoided, how can the programmer, whose ego is involved, really determine whether his nested IFs are complex? The programmer's peers can determine whether the code is complex and will react to it in the same way that a maintenance programmer eventually will. If they think the code obeys the spirit of the standards manual, then it's probably OK; if they don't think so, the fact that it obeys the letter of the standards is irrelevant.

Increasing the readability and overall quality of computer programs (and the models from which computer programs are ultimately developed— e.g., dataflow diagrams, entity-relationship diagrams, and structure charts) are tangible benefits; on that basis alone, organizations have enthusiastically endorsed the walkthrough approach. But there are intangible benefits as well: *training* and *insurance* are probably the two most important.

It doesn't require much imagination to see that a peer-group review can be an excellent way of communicating new ideas or advanced techniques to members of an MIS staff. Indeed, it is surprising that so many programmers and systems analysts spend their entire careers without any significant exchange of new ideas and techniques with their colleagues.

Naturally, we would expect that junior staff members would profit most from a walkthrough approach, because of the opportunity to learn from senior technicians. Sometimes, though, the opposite happens: Senior people who have gotten into a rut learn refreshing new ways of approaching a problem from junior people.[3] Although this can be a source of

---

[3] Note that this is a minor example of a much broader sociological phenomenon, first articulated by anthropologist Margaret Mead. For several thousand years, mankind lived in a "post-figurative" society, in which the younger generation invariably learned from the older generation; the generations that grew up in the early and middle part of the 20th century were a "co-figurative" society, in which parents and children were often forced to learn about new technologies and new trends at the same time. Today we live in a "pre-figurative" society, in which the older generation is often forced to learn from the younger generation. This is particularly true in the computer field, with its rapid technological change; a "generation" in the computer field, as we know, is usually only three to five years in duration. Hence, it is quite common for a recent

embarrassment (a matter to which I will return in a later chapter), it is normally beneficial to everyone concerned.

Finally, there is the insurance factor. In many development projects, an individual programmer's/analyst's work is thrown away if he leaves before finishing his part of the project; the person assigned to take over the incomplete work complains that it's disorganized, undocumented, and thoroughly confusing, and argues that it would be much better to scrap the whole thing and start over. However, this is not true if the work has been subjected to several walkthroughs. In a walkthrough environment, other people would be familiar with the partially completed program; and the peer group would ensure that the work, though incomplete, is designed, documented, and organized properly so that others can understand it. The result almost always is that neither the work nor the time will be lost if the author has to leave the project before its completion.

Thus, the advantages of walkthroughs are many: higher reliability, increased maintainability, better dissemination of technical information among technicians, and an increased likelihood that work can be salvaged if someone leaves the project. In Chapter 10, we will examine some statistics that quantify the effect of these benefits.

1.4   What This Book Will Do for You

This book has three objectives. The first is to acquaint you with the concept of walkthroughs and the related concept of teams. In Chapter 2, we discuss different types of walkthroughs and show where each type fits into a typical development life cycle for data processing projects. In Chapter 3, we discuss the concept of programming teams, sometimes known as "egoless teams," and their relationship to structured walkthroughs.

A second and more important objective is to provide you with guidelines and procedures for successful implementation

---

(3 cont.) college graduate to have more experience with new hardware or software than a ten-year veteran who has been stuck maintaining an old system in the back corners of his organization.

of walkthroughs in your organization. Chapters 4 through 7 discuss roles of different people in a walkthrough, activities that should be carried out before a walkthrough, the conduct of the walkthrough itself, and the follow-up tasks required after the walkthrough.

A third objective of the book is to acquaint you with the psychological problems you are likely to encounter with other technicians and with data processing management during a walkthrough. Peer group reviews are not always as simple, friendly, and objective as one would hope; and programmers and systems analysts are not always as rational, even-tempered, and good-natured as one would hope. The situation is further complicated by MIS management, which is concerned with the impact that walkthroughs will have on the organization. Things can become even more complicated in large organizations, where auditors, users, subcontractors, quality assurance specialists, and others become involved in the development of an information system.

Chapters 8 and 9 are about the psychology of walkthroughs. These two chapters were written originally for programmers, analysts, and other technicians, but since end users are playing an increasingly active role in systems development projects, they should also read these chapters to gain an appreciation of the problems technicians face.

Conversely, Chapters 10 through 13 were written primarily for project managers, but it wouldn't hurt technicians to read them. The more both groups understand about their respective roles, responsibilities, and problems, the better the chances are for the successful implementation of walkthroughs.

1.5   What Assumptions This Book Makes About You

This book assumes a reasonable understanding of computers and information systems development. While I will not dwell on the details of any particular vendor's hardware or any particular programming language, I assume that you are familiar with the concepts of programming, systems design, and systems analysis. I am assuming that the book will be read primarily by programmers, systems analysts, and their immediate managers; however, it should be equally useful for the non-technical end user who has become involved in the development of an information system.

I also assume that you have been exposed to the basic concepts of structured programming, structured design, and/or structured analysis. This assumption is fairly safe, since walkthroughs have been introduced into many organizations as part of an overall package that includes structured programming, structured design, structured analysis, top-down implementation, chief programmer teams, information modeling, and other modern systems development techniques. It's also a necessary assumption because in many cases a walkthrough is practical only if the product has been developed in an orderly, comprehensible way: A proper walkthrough of a large, monolithic "rat's nest" program or a 3,000-page monolithic "Victorian novel" functional specification is almost impossible.

Thus, you should be familiar with the concept of a dataflow diagram, as shown in Figure 1.1; an entity-relationship diagram, as shown in Figure 1.2; a structure chart, as shown in Figure 1.3; and the concept of a project life cycle, as shown in Figure 1.4, which integrates the use of structured analysis and structured design in a typical systems development project. If you are not familiar with these modeling tools, some background reading may be appropriate. Consult [Page-Jones, 1988], [Yourdon, 1988], [DeMarco, 1978], [DeMarco, 1979], [Martin and McClure, 1985], [Orr, 1977], [Orr, 1981], [Higgins, 1979], [Jackson, 1983], [Hansen, 1984], or [Yourdon and Constantine, 1978] in the bibliography for more details.

There is one final assumption. I assume that you are not the only person in your organization who is reading this book—because it doesn't make sense to have a walkthrough if only one person is participating. If your entire group is going to practice walkthroughs, each member should begin with the same vocabulary, concepts, procedures, and ideas about how to avoid the "real-world" problems that will arise.

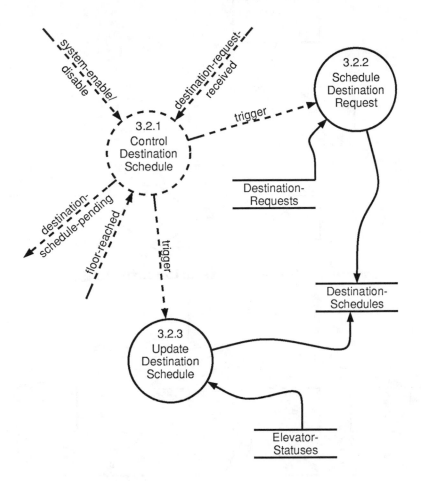

Figure 1.1: **A dataflow diagram**

The questions at the end of each chapter were developed with this in mind. They are not intended to test your ability to memorize the contents of each chapter, but rather to stimulate discussion with your colleagues about

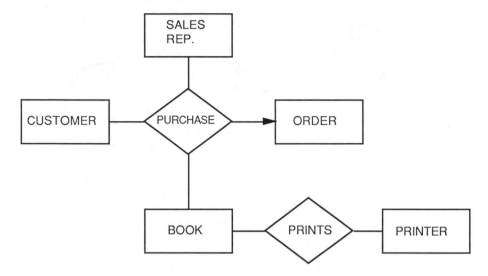

Figure 1.2: **An entity-relationship diagram**

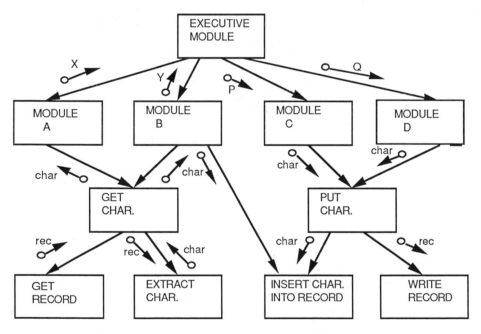

Figure 1.3: **A structure chart**

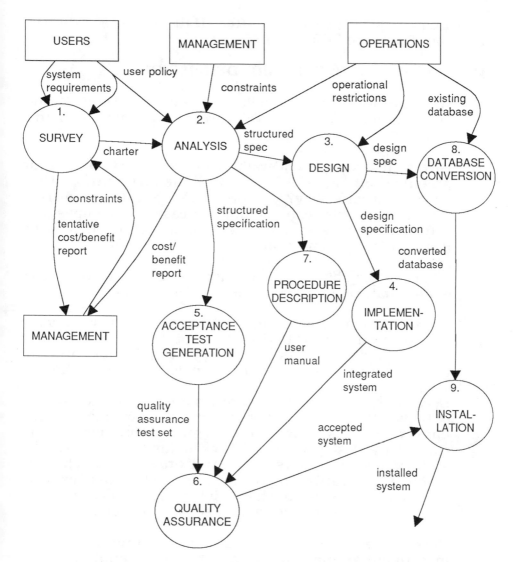

Figure 1.4: **The structured project life cycle**

issues for which the book does not, and cannot, provide black-and-white answers. They are all real questions which have already been raised in many organizations. If you can't find anyone in your own organization who is willing to discuss them with you, write out your answers to the questions (or any new questions you might have) and send them to me at

Yourdon Press, care of Prentice Hall, College Editorial Division, Englewood Cliffs, NJ 07632.

QUESTIONS FOR REVIEW AND DISCUSSION

1. What are walkthroughs called in your organization? How do they differ from the walkthroughs discussed in this chapter?

2. Give some examples of walkthroughs or reviews in your organization outside the MIS department. How successful are they? If you work in an engineering organization, this search should be easy to accomplish; in a typical business-oriented MIS organization, it may be more difficult to find examples.

3. What are the objectives of a walkthrough? Can you think of any that were not mentioned in this chapter?

4. Do the programmers and systems analysts in your organization object to the idea of having their work reviewed by their peers? Do you? What is the nature of the objections? Do you think the objections are valid?

5. Does your organization keep any statistics on the number of bugs found in production programs? How many of the bugs were caused by coding errors? How many were caused by design errors? How many were caused by systems analysis errors? Can you draw any conclusions about the relative importance of code walkthroughs versus design walkthroughs versus analysis walkthroughs? This is a crucial area, as we will see in Chapter 7.

6. How many bugs would you have to find before the investment of time in a walkthrough would be cost-effective?

7. How much of your organization's budget is claimed to be spent on the maintenance of computer systems? What fraction of this figure do you estimate is caused by poor systems analysis, design, and programming when the system was first developed? How much of this could be

saved by having walkthroughs before the system is put into production?

8. Do you think walkthroughs would increase the productivity of the MIS staff? Why? By how much?

9. Does your organization have standards for the analysis, design, and implementation of computer systems? If so, are the standards followed meticulously? At the present time, how can your organization ensure that the programmers are living up to the spirit of the standards and not just to the letter of the standards?

10. What procedures currently exist to train junior members of your MIS staff, and to upgrade the skills of the veteran members? Do you think walkthroughs would be a useful adjunct to these procedures? Why?

11. What are the chances that a partially completed program will have to be scrapped if the programmers have to leave the project suddenly? Do the programmers in your organization have the same opinion as management? Do you think walkthroughs would improve the chances of salvaging partially completed tasks?

12. Do you anticipate any other advantages of walkthroughs? What impact, for example, would they have on the overall morale of the MIS staff?

2 TYPES OF WALKTHROUGHS

> They [the Americans] have all a lively faith in the perfectibility of man, they judge that the diffusion of knowledge must necessarily be advantageous, and the consequences of ignorance fatal; they all consider society as a body in a state of improvement, humanity as a changing scene, in which nothing is, or ought to be, permanent; and they admit what appears to them today to be good, may be superseded by something better tomorrow.
>
> Alexis de Tocqueville
> *Democracy in America, Part I*, Chapter 3
> (1835)

As pointed out in Chapter 1, a walkthrough is a peer group review of some product. Thus, it is meaningful to talk about a walkthrough of anything by anybody at any time. For the purposes of this book, I am interested in discussing only walkthroughs of computer programs and the development of products from which information systems are built. Nevertheless, there is still an enormous amount of flexibility in the types of walkthroughs you may wish to conduct in your organization.

Two major variables determine the nature of the walkthrough. The first of these is formality: How organized, how "structured" should the walkthrough be? The second major variable is timing: At what stage in the development of a product should a walkthrough take place? As we will see in the remainder of this chapter, walkthroughs can occur after the user requirements have been developed, after the design has been completed, after the code has been written, or after the test data has been developed. In many organizations, walkthroughs are conducted at *each* of these milestones in the systems development life cycle.

Walkthroughs can also be used before and during the development of user requirements, design, code, and test data. In addition to verifying the correctness and the quality of the finished product (whether specification, design, or code), walkthroughs can also be used profitably to help develop the product.

2.1   How Formal Should a Walkthrough Be?

Unfortunately, many programmers and systems analysts have the impression that there can be only one kind of walkthrough.   Some visualize the walkthrough as an impromptu bull session in which people gather around a blackboard to argue about a dataflow diagram that has been hastily scribbled in chalk. Others see it as a formal event, to which formal invitations are issued, and for which a formal presentation on overhead transparencies is required. Each group thinks that only its approach is a "true" walkthrough, that one can have either a formal approach or an informal approach but not both.   There can even be "semi-formal" walkthroughs in which the participants decide for themselves just how formal and structured they will be.

We'll begin with the formal walkthrough, since it's the type most commonly known in organizations. It's often called a review or design presentation and is intended to give reviewers a chance to offer formal approval or disapproval of the product.   Often, the reviewers include the developer's boss, and his boss, and his boss's boss, and other "big-wigs," as well as total strangers from the Quality Control Department, the Standards Department, and the user's organization.   In any case, it usually does *not* consist solely of the developer's peers, so it probably should not really be called a walkthrough.

What happens in such reviews depends on the participants and on the product under review. In some cases, the review is nothing more than a rubber stamp, since neither the boss, nor his boss, nor the person from Quality Control really understands the product. In other cases, it turns into a political battle: The user complains that he never wanted the system in the first place, the Standards representative points out that the document violates every known standard in the organization, and the developer's boss blandly says that he had nothing to do with the product and that it's all the developer's fault. And sometimes—but only sometimes—the review turns out to be a frank, honest, productive exchange of ideas, suggestions, and constructive criticism about the product.

While it may not be entirely fair to generalize, many formal reviews are characterized by the following, which I have prioritized in decreasing order of importance:

1. *Slow feedback:* The designer may have to wait weeks between formal reviews, only to find that his past week's work is judged unacceptable.

2. *A general absence on the part of the reviewers of a sense of responsibility for the correctness and quality of the product being reviewed.* The producer usually is viewed as being "stuck" with the product, and the reviewers have no real incentive to ensure that the product is of the highest possible standard before they approve it.

3. *Critiques whose quality is highly variable:* These depend on the competence, the mood, and the attention span of those who attend—as well as the overall political climate in which the walkthrough takes place.

4. *A long preparation time:* Formal invitations have to be issued, the review has to be scheduled to ensure that everyone can attend, and the producer has to spend a great deal of time developing notes, slides, and flipcharts for his presentation.

5. *Relatively complete documentation:* The producer *must* provide such documentation, to avoid embarrassing himself in front of his boss and other important people.

In contrast, the informal walkthrough can be held with relatively little preparation. It can consist of a few people looking at scribblings on the back of an envelope. Consequently, the feedback from such walkthroughs is quick, and the producer often can arrange to have several walkthroughs within a day or two. Because the documentation *is* sometimes written on the back of an envelope, it tends to be much less precise; as a result, the reviewers may overlook errors and flaws. However, the overall quality of the review is often better because the reviewers are peers of the author and presumably are somewhat familiar with the type of work he is doing; in any case, the quality of the review is likely to be more *consistent*. In addition, the producer's peers usually have more of a sense of responsibility because they have to work with the producer daily.

Finally, the semi-formal walkthrough, which I recommend, is informal in the sense that it involves only peers and is composed almost entirely of people who are directly involved in the project on a full-time basis. Yet it is formal in the sense that it has an agenda, a time limit, and an agreed-upon set of procedures for conducting the meeting.

Of course, sometimes informal walkthroughs are thoroughly superficial, and sometimes formal walkthroughs are extremely productive. For a system of any reasonable size or complexity, I recommend all three types of walkthroughs: informal walkthroughs to give the producer a chance to bounce ideas off his peers and see if there are any obvious problems; semi-formal reviews to look more carefully at the basic soundness of the design; and formal reviews as a final check for subtle errors, design flaws, and potential maintenance problems that may have been ignored in the excitement of creating the product.

2.2   When Should a Walkthrough Be Held?

In the previous section, I observed that a walkthrough could be either formal, informal, or any shade of semi-formal. For obvious reasons, informal reviews tend to take place earlier in the development of a program or system, while formal reviews take place later.

To illustrate, let's examine one phase of the development of a program: the coding phase. A code walkthrough could easily take place at any one of the following six stages:

- before the source code has been entered into the computer

- after the source code has been entered, but before it has been compiled

- after the first compilation

- after the first "clean" compilation, free of syntax errors

- after the first test case has been executed successfully

- after the developer thinks that *all* test cases have been executed successfully

Of course, hardly anyone today uses coding sheets to prepare their computer programs. But if we substitute *analysis* for *coding*, these steps are very common: Systems analysts draw dataflow diagrams by hand, then enter them into a CASE system, then check them for errors, etc.

There are advantages and disadvantages to having walkthroughs at these various points in time. For example, it is relatively unpleasant conducting a walkthrough when the source document is a coding sheet because (a) each reviewer probably has a nearly illegible reproduction of a coding sheet whose instructions were written in pencil, with erasures, crossed-out lines, and cryptic notes and insertions; (b) each sheet of paper probably contains only 10 or 20 lines of information, so that the reviewers are constantly turning pages back and forth to see what the program is doing; and (c) the reviewers sorely miss symbol tables, cross-reference listings, and other helpful aids that the compiler produces.

One could wait until after the programmer entered his source program into the computer (most often, these days, by typing his program on a time-sharing terminal, or on a PC workstation), but *before* it has been compiled. Where program compilation is time-consuming, such as when it is done as a third-shift, middle-of-the-night activity, it makes sense to have the walkthrough before the program is compiled. In most organizations today, though, the programmer can arrange to compile his program immediately after entering the source code into the computer, with a delay of only a few seconds before the compilation is complete.[1]

Thus, it is much more common to conduct a walkthrough *after* the program has been compiled. Then reviewers can work with a more legible document with more information on each page, and with the symbol tables, cross-reference listings, and other helpful information which the compiler provides as a matter of course. Some people argue that the walkthrough should not take place until the programmer has produced a clean compilation— that is, one without any syntax errors. In most cases, this makes sense if the MIS organization has reasonably good turnaround for compilations, or if the programmer is working on his own personal workstation. Obviously, the developer must strike a balance. On the one hand, he doesn't want to waste the time of his fellow reviewers looking for syntax errors that can be found in a matter of microseconds by a compiler. On the other hand, he shouldn't spend days repeatedly submitting his program for compilation in an attempt to get rid of difficult syntax errors that he doesn't fully understand.

To delay the walkthrough until the programmer has begun testing his program is usually a bad idea, and it is *definitely* a bad idea to wait until the programmer thinks that he has finished *all* of his testing. First, much time is wasted

---

1 There are still exceptions to this fast turnaround environment. The compilation of a small module may, for example, require access to many large libraries of project-wide symbol definitions, and the compiler may perform exhaustive (and therefore slow) error-checking. A project team at Magnavox found, for example, that it required a full *week* of dedicated computer time on a cluster of powerful VAX computers to fully recompile a system of 750,000 lines of Ada for a system being built for the U.S. Army.

in this kind of approach. The programmer may have spent days looking for his own bugs when a group of reviewers would probably have spotted them more quickly.

Second, there are some ego problems if the developer waits too long to have a walkthrough. I'll discuss ego problems in much more detail in Chapter 9, but I can summarize one of the problems here: If a reviewer suggests, during the walkthrough, that the code should be revised to make it more readable, the producer is likely to become defensive. After all, he has already invested a tremendous amount of time and energy— psychic energy as well as physical energy— and he's not terribly interested in suggestions about rewriting the program, because it will seem that he has wasted the time already invested.

In addition, there is a psychological effect on the reviewers: Finding bugs is one of the things that makes a walkthrough worthwhile for them. In fact, many organizations argue that the more bugs found in a walkthrough, the more successful the walkthrough. On a personal level, a reviewer probably won't mind spending an hour reviewing someone else's code if he finds a bug or two because he feels that his time has been well spent. However, if he spends an hour reading through the code and does *not* find a bug, he thinks that he's wasting his time. He begins to become sloppy, assuming that there won't be any bugs in the code he reviews in subsequent walkthroughs.[2]

The moral is that you can have walkthroughs at any stage in the product's development, but you'll usually find that it is less productive if you have the walkthrough too early or too late. Schedule the walkthrough late enough so that the product is sufficiently well developed and well documented to make some sense, and yet early enough so that the producer doesn't invest too much of his ego in the product, and the reviewers will be able to find some bugs and suggest some improvements. This balance also applies for walkthroughs of

---

[2] My colleague Roland Racko suggests one way of ameliorating this problem: The developer could secretly seed his product with known errors, in order to keep the reviewers awake and interested. Such a process, he says, could be called "defensive bugging."

products other than code; I will discuss other types of walkthroughs below.

2.3   Types of Walkthroughs

In addition to the coding phase, there are many other stages in the development of a typical computer system: systems analysis, design, and testing are among the more obvious ones. Each of these activities can and should have walkthroughs as well.

Specification walkthroughs are, as the name implies, a review of the user requirements, or specifications, of an information system. The walkthrough usually involves the authoring systems analyst, a user representative, and one or more systems analysts on the project. Its main purpose is to spot problems, inaccuracies, ambiguities, and omissions in the system specification. This walkthrough should review the standard components of a structured specification: a set of leveled dataflow diagrams, entity-relationship diagrams, data dictionary, process specifications, and (if appropriate) state-transition diagrams.

The specification walkthrough is concerned with the *functionality* of the system; thus, it deals only with the *essential model* of the proposed system.[3] A second, entirely separate type of walkthrough is known as the *user interface walkthrough*; this is a review of the so-called "user implementation model" of the system. This walkthrough is concerned with input formats and layouts (including the placement of data fields and error messages on screens for an on-line system); output formats and report layouts; and the *sequence* of on-line menus and screens required to accomplish each functional activity. This is also the area where alternative input/output technology needs to be reviewed with the user— e.g., the use of voice input/output, mouse input, digitizers and scanners, etc. All of these may

---

[3] For more about the distinction between essential models and implementation models, see [Yourdon, 1988], [McMenamin and Palmer, 1984], or [Ward and Mellor, 1985]. The essential model is often referred to as the "perfect technology" model: It is a description of the functions and stored memory required of the system regardless of the technology used to implement it.

have been carefully prescribed and documented in earlier phases of the project, but until a real walkthrough takes place, no one will know for sure that the user interface is acceptable. Of course, it is precisely this kind of walkthrough for which prototyping tools and fourth generation languages are so well suited.

Design walkthroughs assume that the functional requirements and user implementation model of the system are correct. The emphasis in a design walkthrough is on the implementation of the system— implementation decisions and strategies that will be "transparent" to the user, as opposed to the implementation decisions like on-line screen formats that are very clearly visible to the user. Design walkthroughs come in many flavors, including the following:

- A "systems level" or "processor level" design, usually documented with augmented dataflow diagrams and supporting documents. It is at this level of design that we decide how to allocate the essential model of the system to different processors and different data storage units; inevitably, this introduces issues of communications networks, distributed systems, etc. A variety of architectural issues can be reviewed by the walkthrough group, including performance, security, cost, and reliability.

- A "subsystem" design, usually documented with structure charts and appropriate supporting documents (data dictionaries, pseudocode, etc.). In a large system, one would find many different walkthroughs of different aspects of the subsystem-level design— for example, walkthroughs of the database design, of the design of the telecommunications subsystem, of the backup/recovery sub-system. The walkthrough group would also examine the manner in which the essential model assigned to each individual processor had been partitioned into different "tasks" or "run units" within the processor. Since inter-task communication almost always takes place through the vendor-supplied operating system, the same issues of performance, security, cost, and reliability must often be reviewed again.

- A "procedural" design, the low-level flowchart or pseudocode of the logic within individual modules. Such a design would immediately precede the actual writing of code for the module.

Code walkthroughs often attract the most attention in organizations simply because code used to be private, and because code is the final tangible product of the development project. The product being reviewed is the code: a program listing. Unfortunately, code walkthroughs sometimes uncover analysis or design problems (usually because the preceding walkthroughs were ignored or performed in a superficial manner), to the dismay of the programmer who spent days writing brilliant code. Conversely, as Glen Myers points out in [Myers, 1979], code walkthroughs often miss a large number of analysis errors for two reasons: The programmers involved in the code walkthrough are unaware of the user's real needs; and everyone assumes that the user requirements were documented correctly and that those requirements were correctly represented in the design. Because of this problem, it is important to note that MIS organizations that conducted code walkthroughs without design and analysis walkthroughs often had mediocre results, and were unimpressed with the concept of walkthroughs.

Test walkthroughs are conducted to ensure the adequacy of the test data for the system, not to examine the output from the test run. Typical attendees include the person who developed the test data, other programmers (including, probably, the author of the program[4]), a systems analyst, and a user representative.

A number of other people can make a useful contribution to *all* of the types of walkthroughs mentioned above. Auditors, Quality Assurance representatives, EDP Operations personnel, representatives of MIS standards

---

[4] Note the strong implication here that the author of the program should *not* be the one who develops the test data.

organizations, and others can help point out serious problems *before* the system is in production. In many organizations, such problems are discovered too late, after the system has been declared operational. The user cannot afford to have the system taken out of operation, and the development staff has already moved on to another project. I will discuss this problem in more detail in Chapter 4.

QUESTIONS FOR REVIEW AND DISCUSSION

1. Discuss the characteristics of *formal* reviews in your organization. Who attends? How quickly can they be set up? How thorough is the review? How much of the review consists of "politics"?

2. What are the characteristics of an informal walkthrough?

3. Do you think that formal walkthroughs and informal walkthroughs can co-exist in your organization? What procedures should there be for determining when a product is ready for a formal review?

4. For a code walkthrough, when do you think the walkthrough should be held—e.g., before the code has been entered into the computer, after it has been compiled, etc.? How much of this depends on the environment in which your programs are developed? For example, would your opinion be different if every programmer had his own workstation, complete with the ability for local compiling and unit testing?

5. Some people argue that even if computer time is cheap and easily available, it is far better to find bugs *before* the program has been entered into the computer. Do you agree? (For more information on this approach, see [Mills, Linger, and Hevner, 1986].)

6. Can you think of any additional types of walkthroughs besides the ones listed in Section 2.3 of this chapter?

What about documentation walkthroughs? Walkthroughs of the conversion from an old system to a new system? Installation walkthroughs?

3 PROGRAMMING TEAMS

Our military forces are one team—in the game to win
regardless of who carries the ball. This is no time for "fancy
dans" who won't hit the line with all they have on every play,
unless they can call the signals. Each player on this team—
whether he shines in the spotlight of the backfield or eats dirt
in the line—must be an All-American.

General Omar Bradley
Testimony to the Committee on Armed Services,
House of Representatives, October 19, 1949

Before proceeding with a discussion of walkthroughs,
we need to discuss briefly the related topic of programming
teams. The original idea of walkthroughs was suggested
largely by Gerald Weinberg in his classic book, *The Psychology
of Computer Programming* [Weinberg, 1971]. A more recent,
and superbly written discussion of the "peopleware" issues in
systems development, including programming teams, can be
found in [DeMarco and Lister, 1987]. Indeed, whenever you
hear discussions about walkthroughs, you are likely to hear
such phrases as egoless programming, programming teams,
democratic teams, adaptive teams, and even programming
families. While your present organization may never
implement the concept of programming teams, it's a good
idea to understand what they are; it may be something for you
to look for in your next job interview.

3.1   Background of Teams

In both social and industrial organizations, the value of teams has been recognized for hundreds—perhaps thousands—of years. In the past decade, for example, several automobile companies have experimented with the idea of teams that are responsible for the complete production of whole subsystems of a car. As an alternative to the traditional assembly line approach, the team approach generally has led to higher productivity, lower absenteeism, higher quality, and improved employee morale. The team normally schedules its own daily work assignments and even handles its own minor disputes and discipline problems. All of this is usually "invisible" to management, which is concerned primarily with the overall productivity and quality of the work. This concept has received considerable attention in the past few years because of the tremendous successes reported in Japan and other countries that have used the concept in manufacturing industries; see, for example, [Lewis, 1978] and [Ouchi, 1982].

As another example, consider the teamwork found in championship football, baseball, and basketball teams. While individual players may make the headlines because of a spectacular play or a streak of outstanding performances, the team depends on the talents of all of its members for long-term success. Indeed, there are many instances of a team losing a game in spite of the outstanding performance of one of its members.

I could devote a considerable amount of discussion to the psychology and management of such teams, but that is not the purpose of this book. For general information on the formation and management of teams in business organizations, consult [Aukee, Beckwith, and Buttenmiller, 1973], [Dyer, 1977], [Francis and Young, 1979], [Drury, 1984], [Ends and Page, 1984], [Rimler and Humphreys, 1980], [Ford, 1983], [McCulloch, 1982], and [Bass, 1975].

Excellent discussions about the formation and management of *programming* teams are provided by [Aron, 1983], [Semprevivo, 1981], [Thomsett, 1981], [Weinberg, 1971], and [DeMarco and Lister, 1987]. If you are serious about implementing true programming teams, you should study these books carefully. The main point is that to have a team of cooperative peers, whose minute-to-minute activities are decided without the intervention of management, is not such a radical idea.

3.2   A Brief Look at Classical MIS Organizations

Now compare the structure and dynamics of the teams mentioned above with those of the classical MIS organization. In most such organizations, a manager, often with the title of project leader or lead programmer, determines the work assignments and delegates individual programming tasks to the programmers. If the project is a smashing success, usually the manager receives the largest share of glory; if the project fails or falls behind schedule, the project manager also bears the brunt of the criticism from upper management. Consequently, we often find that the programmers feel no real responsibility— other than their own private compulsions— to make the project a success. They have no incentive to share their programs with their peers or to invest any significant energy in helping their colleagues find bugs.

In addition, the situation is frequently cold-blooded. In some organizations, programmers know that a limited amount of money is available at year's end for salary increases; concurrently, they know that only one or two slots are available for promotion to higher grades within the organization. If there are ten programmers and only two slots for promotion, competition among programmers exists. Each programmer may feel that if he spends too much time helping his colleagues, it will detract from his chances of getting a raise or a promotion.

In such an environment, it is common for each programmer to develop an ego attachment to his programs. Indeed, he refers to them as *his* programs, his personal property. At the very least, the programmer is defensive about his code and is reluctant to accept any suggestions or constructive criticisms from other programmers. In the extreme, he hides his listings and refuses to show the programs to anyone. The situation can be even more bizarre, with some systems analysts viewing the system requirements being documented as *their* requirements rather than those of the end user—an attitude the user has great difficulty understanding.

As an example, read what programmer Bill Atkinson had to say in [Goodman, 1987] about the development of the original QuickDraw program for the Apple Macintosh computer:

> ...when I was working on QuickDraw, that was all mine. I wrote it all myself. I didn't let anybody look at my sources. I listened to what people needed in the labs, but it was my call on everything. Then I went from that to MacPaint, which was more of a shared thing, because I would accept a lot of kibitzing from people who could use it as I was developing it. Still, I wrote every line of code. Nobody saw the source until after we had shipped.

As we will see in Chapter 14, Atkinson's approach changed by the time he began developing the highly successful HyperCard system for the Macintosh a few years later.

3.3   The Objectives of Programming Teams

This kind of environment sounds pretty grim, and maybe it is a bit extreme. Unfortunately, members of large MIS organizations—such as banks, insurance companies, and government agencies with upward of a thousand programmers and systems analysts—are beginning to behave more and more as described above. In contrast, programming teams are an attempt to create a more human, more productive programming environment; it is interesting to note that such teams are more visible in many of the smaller software firms and PC companies.

The purpose of teams in the programming field is to change systems development from "private art to public practice," as Harlan Mills puts it. Or, to use Weinberg's words, the intention is to change programs from "private works of art to corporate assets."[1] That is, instead of creating

---

[1] If this concept is applied diligently, it can help an MIS organization develop libraries of *reusable modules*, so that new systems can be constructed largely by "fabricating"

an environment where programmers work alone throughout their entire careers, with programming teams we create an environment in which everyone feels free to discuss and critique everyone else's programs. Why? Primarily because the team is organized so that its members have a common goal: Each member can expect similar rewards if the project succeeds and similar penalties if the project fails (just as in a baseball team, where the whole team suffers if one person strikes out at a critical point, and where several players working in harmony are needed to carry out a critical double play).

One of the benefits of the team approach is that it can take advantage of the talents of different members at different stages in the project. Just as a baseball team depends on good pitching at certain points in the game, good fielding at other stages, and strong hitting at still others, so a project team finds it needs varying degrees of expertise in systems analysis, design, programming, testing, and debugging at various stages of the project. Some people are excellent systems analysts but are not familiar with the hardware and software technology that will be used to implement the system. Others are good at design, but unimpressive when it comes to the dirty work of grinding out the code; still others are unimaginative at the design stage, but are workhorses who can produce phenomenal amounts of code once the design has been sketched out. Some have a talent for generating test cases to shake down a system thoroughly, while others are best at the mysterious art of debugging: They can read a 500-page hexadecimal dump and actually "see" bugs lurking among the hexadecimal digits. Forced to work on their own, each of these people will excel in one or two stages of their assignment and will do a mediocre job at every other stage; working together as a team, they can take advantage of critical talents when and where they are needed.

(1 cont.) modules that have been previously written and thoroughly tested. This approach is easier in a technological environment such as that provided by the UNIX operating system, but it also requires a corporate attitude and a personal mindset geared toward developing reusable code. For more on the managerial and cultural problems of developing reusable code, see [Berber and Oktaba, 1987], [Gargaro, 1987], [Scott, 1987], and [Tracz,1987].

As John Sculley, CEO of Apple Computer, eloquently points out in *Odyssey* [Sculley, 1987]:

> The beauty of a network is that it has no center. It is a process more than a structure, composed of modular groups that establish themselves to take on specific tasks—not to build fiefdoms as traditional "departments" do. Depending on the situation, the leader can thus also be a follower and a peer, offering inspiration, not his own dogmatic views. Often I am a leader in one network but a follower in another, as I take a back seat to players who are stronger in product development or manufacturing. The corporate leader is not necessarily a paragon of wisdom: in most second-wave companies he is the end product of a process of elimination, not a process of cultivation where talent and ideas shine.

Chances are that you've already seen examples of teams (albeit informal ones) in your data processing career. One of the best examples is in the training of MIS personnel. Compare the environment in a typical university programming course with the environment in a typical industry training program. In the university, each student is expected to compete because there are only a few A and B grades to be given out. If the instructor finds that a student has been helping a fellow student get his program working, he's likely to flunk both.

In industry, on the other hand, the basic object is to make everyone succeed in the difficult task of learning how to program, design, or carry out systems analysis. After all, the organization has invested a substantial amount of time and money in the fledgling systems development technician, and it is uneconomical to set things up so that people fail if they don't have to. At the same time, the training budget in most MIS organizations is so small that the instructor often doesn't have enough time to give adequate personal attention to each student, so he simply tells the students that they should use the buddy system to help each other. In most such training programs, groups of two or three trainee programmers huddle around a program listing (or huddle around a personal computer or workstation), trying to figure out why one of them is having a problem. In the end, they all learn from the process, and they all win. And so does the organization, until it decides to put them into a "real" systems development project, where programmer-to-programmer competition and other political realities of modern corporate life usually destroy any tendencies toward teamwork that the new programmers may have had.

3.4   How a Team Functions

To some, this description of programming teams sounds highly radical, indeed, almost communistic. In practice, it is rarely anything of the sort.   Members of a typical programming team are law-abiding, reasonable human beings who look, act, and think just like anyone else.

In fact, organizations sometimes have programming teams without even knowing that they have teams. The most common example is a typical development project where the manager is inexperienced, uninterested, or unable to provide strong leadership, and the technicians gradually discover that they have common ideas and interests in the analysis, design, and implementation of their system.   In such a case, the technicians, as well as management, may be unaware that they actually are operating as a team.

In organizations where teams have become a formal concept, it is still common to see a project used as the justification for forming a team.   The team is given an assignment, with appropriate constraints (budget, schedule, etc.), and is then given considerable freedom to develop its own day-to-day procedures.   The critical point is that the team operates in an environment where all of the team members feel free to discuss and critique each other's work. This discussion of other people's programs is usually formalized in the *walkthrough* approach.

3.5   Problems with Teams

The comments above imply that programming teams are neither very common nor very popular. Indeed, the notion of programming teams does run counter to the classical way of organizing programmers on a project.   Why?   What kind of problems should you expect in a programming team?

One characteristic of the true programming team is that nobody is really in charge or the boss in the traditional sense. This feature makes outside management somewhat nervous ("Whose rear end are we going to kick if the project comes in behind schedule?"), although that problem can sometimes be circumvented by nominating one team member as the spokesperson for the group (better still if it can be a rotating responsibility). The absence of a formal boss may also be difficult for some team members to handle; many people are accustomed to, and would prefer, an authoritarian manager to tell them what to do.

Note also that the typical programming team probably doesn't have any superprogrammers. If it did, we would call it a chief programmer team in keeping with the terminology IBM introduced. For more details, see the classic paper by Terry Baker of IBM in [Yourdon, 1979]. However, most companies have no superprogrammers anyway, so perhaps the programming team is the best way to cope with the large number of average programmers a company employs.

Obviously, the mere act of putting three or four people into a project does not make them function automatically as a team. Implicit in the concept of a team is the idea of working closely together, reading each other's work, sharing responsibilities, getting to know each other's idiosyncrasies (both on a technical and personal level), and accepting group responsibility for the product. If this attitude can be instilled, the effect is usually one of synergism: Five people working together on a team may produce twice as much as they would working individually.

There is another difference between the chief programmer teams and the egoless teams. If you have a superprogrammer in your organization, chances are that he's not egoless—he's very good, and he doesn't hesitate to tell everyone (including the user, who generally doesn't care) just how good he is. This personality may well destroy the democratic spirit of the team, especially when it comes to walkthroughs. You may find that the superprogrammer (or chief programmer) wants to review all the work *by himself*, rather than making it a group activity—and that sort of one-

on-one confrontation between an individual program-
mer/analyst and the superprogrammer is hardly likely to be
egoless.

Many people feel that programming teams are the wave
of the future. In organizations where the concept has been
implemented successfully, the results have been quite
impressive and the programmers maintain that they will
never revert to their old way of doing things. However, some
organizations have found that they simply cannot implement
the concept; there are too many psychological problems, too
many personality clashes, and too many political problems.

In summary, it seems that the concept of true
programming teams, or families, will be implemented only in
a small number of relatively progressive MIS organizations.
Indeed, we may find that its acceptance parallels the growth
of the "electronic cottage industry," in which programmers
and systems analysts spend part or all of their time working at
home from a terminal or PC. (For more on the cottage
industry, see Chapter 3 of *Nations at Risk* [Yourdon, 1986].)
We may also see programming teams forming in end user
groups, where PCs and fourth generation languages make it
possible for a small group to build information systems.

But if this is the case, why are we discussing teams?
The reason is that about the only part of the team concept
that is likely to be implemented on a wide scale is the
walkthrough. A systems development environment that
encourages walkthroughs is one of the first steps toward
establishing programming teams.

QUESTIONS FOR REVIEW AND DISCUSSION

1. Has everyone in your organization read Weinberg's *The
    Psychology of Computer Programming*? If not, why not?
    If so, make sure they have all read DeMarco and Lister's
    Peopleware as well.

2. Give some examples of teams outside the systems
    development profession. Can you see anything they would
    have in common with programming teams?

3. In your organization, do programmers, designers, and systems analysts openly compete with each other for raises and promotions? If so, does it have any noticeable impact on the way information systems are developed? What impact does it have on morale?

4. Is it common for the programmers in your organization to feel possessive about their programs? What about systems analysts: Do they feel possessive about the system specifications they develop, or about the entire system that they are responsible for specifying? What attitude does MIS management have toward this ego attachment?

5. What do *you* think the objectives of a programming team should be?

6. Do you think that programming teams will work in your organization? Is it likely that everyone will receive a similar reward if the project is successful? Do you think that teams could exist without the tacit approval of MIS management?

7. What problems would you anticipate if programming teams were formed in your organization? How difficult would the problems be to solve?

8. Do you think that walkthroughs can be successfully implemented in an organization that does *not* endorse the concept of programming teams? Why or why not?

4 ROLES IN A WALKTHROUGH

All the world's a stage,
And all the men and women merely players.
They have their exits and their entrances;
And one man in his time plays many parts,
His acts being seven ages. At first the infant,
Mewling and puking in the nurse's arms.
And then the whining school-boy, with his satchel,
And shining morning face, creeping like snail
Unwillingly to school. And then the lover,
Sighing like furnace, with a woful ballad
Made to his mistress' eyebrow. Then a soldier,
Full of strange oaths, and bearded like the pard,
Jealous in honor, sudden and quick in quarrel,
Seeking the bubble reputation
Even in the cannon's mouth. And then the justice,
In fair round belly with good capon lined,
With eyes severe and beard of formal cut,
Full of wise saws and modern instances;
And so he plays his part. The sixth age shifts
Into the lean and slippered pantaloon,
With spectacles on nose and pouch on side,
His youthful hose well saved, a world too wide
For his shrunk shank; and his big manly voice,
Turning again toward childish treble, pipes
And whistles in his sound. Last scene of all,
That ends this strange eventful history,
Is second childishness, and mere oblivion,
Sans teeth, sans eyes, sans taste, sans everything.

William Shakespeare
As You Like It, Act II, vii, 139

A successful walkthrough involves several people, each of whom plays a definite role. The roles need not be permanent: for example, one person can be a coordinator in one walkthrough and a secretary/scribe in the next. Indeed, it is possible and often desirable for one person to play more than one role in a single walkthrough.

The changing of roles is particularly common in small development projects where there may be only two or three people available to participate in the walkthrough. The important thing is to recognize what the significant roles are so that you make sure they are properly played. In subsequent sections of this chapter, we will examine each role in turn. I will mention only briefly the specific duties of each role; the details of those duties will be discussed more thoroughly in Chapters 5, 6, and 7. The roles that we will discuss are:

- The presenter

- The coordinator

- The secretary/scribe

- The maintenance oracle

- The standards bearer

- The user representative

- Other reviewers

4.1   The Presenter

The most obvious role is that of the *presenter*. A typical walkthrough involves at least one person whose job it is to introduce the product—the program listing, the user requirements, or the systems design—to the rest of the group. We will see in Chapters 5, 6, and 7 that the presenter has important tasks before, during, and after the walkthrough.

Most often, the presenter is also the *producer* or *author* of the product under review because it is the producer who most wants to know if there are any problems with the output. It is the producer who will have to fix the flaws or bugs that are detected; and it is the producer who, since he knows his product, is in the best position to present it to others.

However, some people argue that for just this reason, the producer should *not* be the presenter: There is an excellent chance that the producer will brainwash the reviewers into believing that the product "works" just as he thinks it does. As a result, the reviewers may overlook bugs or errors. In addition, as my colleague Nan Matzke points out, the author will often provide verbal explanations that are not supported by the documentation (e.g., "I know it's not evident from reading the data dictionary, but I can tell you the details of this data element.").

To avoid this problem, many organizations ask that the product be presented by someone other than the author; it may be the "buddy" of the author, or it might even be someone who was not directly involved in any aspect of the product's development. For example, if the product is a set of dataflow diagrams that represents the user's functional requirements, perhaps the presentation should be made by the user who will ultimately "own" the system.

We can carry this one step further: In some walkthroughs, there may not be a need for *any* presenter. If the product is sufficiently self-explanatory, the reviewers should be able to study the appropriate documentation on their own, and then attend a walkthrough to discuss their criticisms and suggestions. This is becoming more common in organizations where the systems development models (dataflow diagrams, entity-relationship diagrams, etc.) are produced with automated, PC-based tools, and where the development project members are connected to one another via electronic mail and other office automation support tools.

For the remainder of this book, I will assume that the walkthrough *does* involve a presenter. Local circumstances usually will determine the need for this role, and the rest of the walkthrough approach works the same regardless of whether or not there is a presenter.

4.2   The Coordinator

Perhaps the most important role is that of the *coordinator*. As the name implies, the coordinator ensures that the walkthrough activities are properly planned and organized. The next three chapters will describe the crucial role the coordinator plays both before and after the walkthrough, ensuring that everyone remembers to attend

During the walkthrough, the coordinator serves as a moderator and keeps the discussion on course. It is often suggested that the coordinator should be the team leader, project manager, chief programmer, or some other senior person. While this is possible, I advise against it; Chapter 11 will elaborate on the reasons for avoiding a heavy management

4.3   The Secretary/Scribe

Another important role is that of the *scribe*, or the *secretary*. As the name implies, this person takes the walkthrough notes, which serve as a permanent record and are critically important if the walkthrough is going to be used for ongoing quality assurance of the product.

Some suggest that the scribe can be a clerical secretary—someone who is good at taking shorthand and typing, but probably unfamiliar with the details of programming, systems design, or other technical aspects of the product. In systems analysis walkthroughs, this may not be a problem since the clerical person may be familiar with the business-oriented jargon used. In general, however, the choice turns out to be a mistake. Since the reviewers' comments may be made quickly and in cryptic terms (programming terms, hardware terms, acronyms, and abbreviations that only the project members can understand),

In addition, the scribe needs to know which reviewers' comments can be ignored and which need to be written down. We don't want a verbatim transcript but rather an intelligent summary of errors found, suggestions made, and questions raised. Consequently, I suggest that the scribe be a participating member of the team. In any case, it should be a person who is familiar with the technical details of the project under review.

One other point should be made: The scribe usually will be so occupied with note taking that he will be unable to actively participate in the discussion of the product. Hence, if the scribe also wishes to play a role as a reviewer, it is best that he review the product individually *before* the walkthrough and summarize his comments in writing. Alternatively, the role of scribe could be rotated from person

4.4   The Maintenance Oracle

The roles discussed thus far have not been of a *reviewing* nature. Instead, they were concerned with the presentation and administration of the walkthrough. Obviously, a walkthrough has to have reviewers as well; it is important to recognize that different *types* of reviewers are

One type of reviewer is a person we'll call the *maintenance oracle*. His role is to review the product from the viewpoint of future maintenance. The reason for using the term "oracle" is that the reviewer must try to look into the future, to anticipate what kinds of changes might have to be made to the product. For example, the maintenance oracle should ask himself whether the product is essentially self-documenting, and whether it could be maintained if the producer were no longer in the organization.[1]

Frequently, the maintenance oracle is a person who does not work in the group producing the product. For

---

[1] This is more crucial than you might think. Studies by ICL (the British computer manufacturer) showed that as many as ten generations of maintenance programmers might be involved in patching, debugging, and "tweaking" a program before it finally collapses of old age. As we enter the 1990s, many organizations are maintaining systems that were developed in the early 1970s—or even earlier. Thus, it is quite possible to find third generation maintenance programmers talking to third generation users about a proposed change to a system. Neither the maintenance programmers nor the current users were involved in the original development of the system; indeed, it is possible that they were not even *alive* when the system was first specified!

example, the maintenance oracle may be a member of a Quality Assurance department, or the maintenance department. These departments will have to look at the product sooner or later, and by including such a person in a walkthrough at the early stages of product development, many nasty confrontations can be avoided.

Most maintenance oracles are concerned only with the maintenance of the programs that will eventually operate on a computer; hence they tend to be interested only in code or design walkthroughs. However, it is increasingly important that they attend specification walkthroughs. The computer industry is gradually realizing that a change to an operational computer program should not be allowed until the "statement of requirements" is changed, resulting in the appropriate changes to the design models. *Only then* can the code be changed. If we are to do business in this fashion, as we eventually must, then it becomes crucial for the maintenance oracle to ensure that the specification itself (the dataflow diagrams, the data dictionary, the entity-relationship

4.5   The Standards Bearer

Another important role is that of the *standards bearer*. As the name implies, this role is concerned with adherence to standards— programming standards, design standards, analysis standards, or any other kind of standards that the organization follows. This is an important role because standards need to be interpreted in an "intelligent" fashion: Obeying the letter of the standards is not as important as obeying the spirit. By including a standards bearer in the walkthrough, it assures that someone will speak as a defender of the standards, and counter-comments can be made by

I often find that the standards bearer and the maintenance oracle come from the same department in an organization; indeed, they may even be the same person.

4.6   The User Representative

When appropriate, the walkthrough should also include a *user representative*, someone who can ensure that the product is meeting the customer's needs. Among other things, this step helps avoid the unfortunate phenomenon of creating a brilliant solution for the wrong problem. In most organizations, the user, or a user representative, participates in the earlier product walkthroughs. For example, the user should play a major role in a systems analysis walkthrough; as mentioned earlier, it may even make sense for him to be the presenter in such a walkthrough. The user must also play a major role in the walkthrough of the "user implementation model," where such user-interface issues as screen layouts and report formats are discussed. The user might also participate in design walkthroughs, but probably not in code walkthroughs—unless the users are involved in implementing the system, using a fourth generation language or other high-level system building tools. And, finally, users should definitely be involved in walkthroughs of the system test data.

4.7   Other Reviewers

A typical walkthrough will include one or more additional reviewers whose primary role is to give a general opinion of the correctness and the quality of the product. Such reviewers usually work in the same team or the same project as the producer, though not always. In fact, sometimes it is extremely useful to include one or more reviewers from outside the immediate group. They can bring a fresh, objective perspective to the product, while the

One example of an appropriate "outsider" is the EDP auditor. Many EDP auditors, especially those in the "Big 8" accounting firms, are now familiar with the modeling techniques of structured analysis, structured design, and structured programming. They are learning to distinguish between those audit controls that must be included as part of the "essential requirements" of a system (regardless of the technology used to implement the system), and those controls that should be introduced during the design phase of the project, when alternative technological (hardware/software) choices are being evaluated. For a good discussion of building audit controls into systems, see *Building Controls into Structured Systems* [Brill, 1983].

Another important outsider is the manager of the organization's data center, or a suitable representative from the data center. Indeed, this is even more important if the system is going to be built on a PC and operated within the user's facilities, because most users have little experience with backup, security, audit trails, environmental problems (heat, dust, the impact of donut crumbs on floppy disks, etc.), interactions with outside vendors (e.g., the telephone company), etc. If a new system is going to place major demands on the organization's centralized computing resources, or if new hardware, telecommunication equipment, or software packages must be acquired, the data center manager should be brought into the discussion as early as possible. For a good discussion of the relationship between

Other participants might include the representatives (or "owners") of existing systems with which the new system must interface; computer security specialists; data administrators and/or representatives from the Information Resource group; documentation specialists; and technical writers who can evaluate the readability of user manuals and

4.8   Who Chooses the Reviewers?

In the previous sections, I have identified several useful roles in a walkthrough. That leaves us with one major question: Who decides which people attend the walkthrough? For example, who decides whether it is important to have a maintenance oracle or whether a user representative should be invited?

The answer is...it depends. Based on the project's organization, either the producer will decide who should attend the walkthrough, or the project manager will decide, or the team will decide.

In a team environment, of the sort discussed in Chapter 3, the entire team is responsible for the correctness and quality of the final product. In that case, all the members of the team should attend the walkthrough, and the team should decide whether additional "outsiders" are to be invited.

In a typical non-team environment, each programmer or systems analyst is usually held personally responsible for the correctness and quality of his own products. Consequently, the producer invites whomever he wants. If he deliberately avoids inviting key people, such as the standards bearer, it is he who will eventually suffer.[2] On the other hand, if there are reviewers with whom the producer simply cannot get along, he has the freedom to avoid inviting them. In an organization where management plays a more active role, the project manager or team leader may decide who should attend. For more formal reviews, there may even be organizational standards that dictate who should attend.

Several approaches to selecting reviewers can co-exist in the same project. At the earlier stages of a product, the producer may wish to invite only one or two people just to have a "quick and dirty" walkthrough to see if there are any major flaws. As the product comes closer to completion, the walkthroughs may become more formal, and the selection of reviewers may be increasingly outside the producer's control.

Finally, how many people should attend the walkthrough? As we have seen from the previous discussion, at least six distinct roles can be identified. Of course, some roles may be combined, and in some walkthroughs it may not be necessary to include certain roles. Most organizations find that five to six people are enough for a good walkthrough, and that more than that becomes unproductive. Conversely, a walkthrough can be held by as few as two people (the producer and one reviewer) and will often involve only three or four people. The fewer participants and roles, the more likely that certain bugs and flaws will be overlooked; the more participants (above five or six), the more likely the walkthrough will degenerate into a committee meeting.[3]

---

[2] If you believe in the perspective of Chapter 3, then you would disagree with this statement—for while the producer might suffer first and most grievously because of a bug in his program, ultimately the entire team suffers because the entire system will either be late or bug-ridden.

[3] My colleague Nan Matzke makes an excellent point here: Some products need to be reviewed from several different perspectives, and the situation may call for multiple walkthroughs with different people attending. The user implementation model, for example, needs to be reviewed to ensure that it satisfies the user's wishes concerning

QUESTIONS FOR REVIEW AND DISCUSSION

1. Do you think the producer should present his own product in a walkthrough? What are the advantages and disadvantages of such an approach?

2. How easy do you think it would be for the producer to brainwash the reviewers into believing, as he does, that his program is correct? Can you cite an example?

3. Do you think it's possible for a walkthrough to take place without *any* formal presentation of the product? Under what conditions would this make sense? What are the advantages and disadvantages?

4. Do you think that the coordinator should be a project leader or some kind of management representative? What are your reasons? Does management agree with

5. Do you think the scribe *must* be a programmer or technical person? Could the scribe be a clerical secretary? Could the scribe be a program librarian found in many chief programmer team organizations?

6. Should the maintenance oracle be a member of the same group that is building the system, or should he be an outsider? How is this handled in your organization?

7. How strong a voice should the standards bearer have in the walkthrough? Should he be left out of the early, informal walkthroughs and not included until the product has begun to stabilize? What is the opinion of the standards people in your organization?

8. How many walkthroughs do you think a user should attend? What do the users in your organization think of

9. What kind of additional participants could play a constructive role in a walkthrough?

(3 cont.) input formats, output display formats, etc. But it also needs to be reviewed from a technical implementation perspective, a possibly separate walkthrough which the user may not wish to attend.

10. Who do you think should determine the makeup of the reviewing audience— the producer, management, or the systems development project team? Why? Does it always have to be done the same way?

11. How many people do you think should attend the walkthrough? Do you think there should be a minimum and maximum number of attendees?

5 ACTIVITIES BEFORE THE WALKTHROUGH

> If men are to be precluded from offering their sentiments on a matter which may involve the most serious and alarming consequences that can invite the consideration of mankind, reason is of no use to us; the freedom of speech may be taken away, and dumb and silent we may be led, like sheep to the slaughter.
>
> George Washington
> *Address to officers of the Army,* March 15, 1783

As we saw in Chapters 1 and 2, there are many variations of the basic walkthrough format. One area where there is a great deal of variety is in the amount of preparation that takes place before the walkthrough. At one extreme is the programmer who dashes off the last line of code, or the systems analyst who finishes drawing the last bubble of a dataflow diagram, leaps on top of his desk, and yells to the entire office, "Anybody here wanna have a walkthrough?" At the other extreme is the programmer or systems analyst who treats the walkthrough as if it were a black-tie affair, to which he issues formal, engraved invitations.

Obviously, neither extreme is desirable. But preliminary organization and preparation are necessary to make the walkthrough as effective and productive as possible. Since the time of half a dozen highly paid people is involved, the walkthrough should be run as if every minute is precious—because it is.

This chapter discusses the preparations needed to make the walkthrough move smoothly and briskly, by reviewing the activities of the producer, the coordinator, and the participants. Most of the discussion assumes that the project team is still working in a primitive environment, where documents are scribbled on scraps of paper, reproduced manually on copy machines, and distributed by walking around the office dropping the appropriate scraps of paper on the appropriate desks. It is indeed ironic that so many data processing people do work in such an environment, while the end user community is pushed to move to a modern world of office automation, electronic mail, automatic document filing and distribution, and so forth.

Happily, some MIS organizations are beginning to use these tools for their own work. The introduction of PCs and local area networks in the systems development organization has brought with it electronic mail and associated office automation facilities. And there are now a number of PC-based CASE tools for automating the creation of dataflow diagrams, entity-relationship diagrams, etc.; among these are Index Technology's Excelerator, Nastec's CASE 2000, and YOURDON inc.'s Analyst/Designer Toolkit.

However, as this book is being written in 1988, only about 2 percent of the professional programmer/analyst community in the United States has access to such tools. It is estimated that this will increase to approximately 10 percent by 1990, and 50 percent by the middle of the 1990s. Thus, I will have to wait until the next edition of this book to assume that a majority of systems developers have modern CASE-oriented technology available to them!

An MIS organization that *does* have such facilities available will find that the administrative activities described in this chapter can be accomplished quite easily. If your organization does not fall into this category, you will have to resort to quill pens and parchment.

5.1 Responsibilities of the Producer

The producer initiates the walkthrough. While his responsibilities are obvious and almost mechanical, they are crucial to the success of the walkthrough.

The producer's first responsibility is to announce his intention to have a walkthrough. This notice should be given at least two days in advance, so that the participants will have adequate time to prepare. If the producer is certain that the participants will have ample spare time, the walkthrough can be announced one day in advance or even one hour in advance. However, most MIS organizations are far too hectic to allow for such last-minute planning: An important participant is out of the office, another is in a meeting or in another walkthrough, and yet another is engaged in a critical project that can't be interrupted for a day or two.[1]

The producer's second responsibility is to choose a coordinator and participants for the walkthrough. As I discussed in Chapter 4, some organizations may take this decision out of the producer's hands: The entire team may attend the walkthrough as standard practice, or the project manager may decide who should attend. But it is common for the producer to decide who would be most appropriate to review his program or his systems design.

The producer is also responsible for providing appropriate documents for the walkthrough, such as program listings, structure charts, and dataflow diagrams. These materials are usually given to the coordinator, who arranges for their distribution to the other participants.

What should the producer do to occupy his time during the two days before the walkthrough? In most organizations, this is not a problem: The producer has other programs or specifications to write, other projects to work on, or even other walkthroughs to attend. If he is concerned that he

---

1 Note that in an automated office environment, the walkthrough could be scheduled with an on-line calendar system, which would automatically notify all of the participants via electronic mail.

might not have anything to do, the producer could announce his walkthrough a day or two before he actually finishes his product. However, this would mean that he probably would not have any documents ready for distribution; consequently, the reviewers would be forced to attend the walkthrough cold, and the quality of the walkthrough will suffer accordingly.

It is not easy to choose between having a producer who may spend an idle two days waiting for the walkthrough, or having a group of reviewers who may be wasting their time reviewing a product that they have not seen before. The best possible solution is the one in which the producer and the reviewers can break their assignment into small tasks, each of which can be worked on individually. A project developed using top-down structured analysis, structured design, and structured programming has the advantage of breaking a system into small, highly independent modules, each of which can be worked on separately by one or more analyst/programmers.

The producer has one final responsibility: to choose a product that *can* be reviewed in a period of 30-60 minutes.[2] This restraint necessarily limits the scope and size of the product. One can expect to walk through 50-100 lines of code in a typical third-generation or fourth-generation language; or one to three pages of a structure chart; or one or two dataflow diagrams with accompanying data dictionary definitions. Obviously, if the producer is developing a large product, the walkthroughs have to be done one piece at a time; once again, a top-down structured approach eases this procedure since it organizes a system into a hierarchy of modules, each of which can be reviewed independently.

5.2   Responsibilities of the Coordinator

The coordinator has important responsibilities before, during, and after the walkthrough. The most important of these is to ensure that the walkthrough actually takes place; his other responsibilities ensure that the walkthrough runs smoothly and effectively.

---

[2] This is discussed further in Chapter 6, but the reason is quite simple: The reviewers' attention span and ability to concentrate drop significantly after an hour.

Once the coordinator has been chosen by the producer, he selects a time and area for the walkthrough. He may need to reserve a meeting place such as a conference room, and he must contact each of the other participants to choose a mutually convenient time and place.

The second responsibility of the coordinator is to ensure that the participants attend. In most cases, the coordinator has to contact personally each of the participants and confirm that they will be available for the walkthrough and that they have the time and interest to devote an hour to reviewing the producer's documentation prior to the walkthrough. Depending on the situation, the coordinator may have to play the role of mother hen— coaxing, nagging, and chiding the participants to attend the walkthrough.

Last, the coordinator is responsible for distributing the documentation for the walkthrough. He usually must reproduce several pages of material as well as distribute them to the participants. As mentioned previously, these activities can be done easily with appropriate office automation equipment; without it, the coordinator either has to do them himself or delegate them to a clerk.

5.3   Responsibilities of the Participants

Finally, we come to the participants. Their primary responsibility is simply to agree to be participants and to contribute the necessary time and energy to make the walkthrough successful. While this is an obvious point, it needs to be emphasized: *If a participant feels that he will be unable to devote enough time to review the product, it is better that he not participate, rather than perform a superficial review.*

How much time is necessary depends on the nature of the product under review and on the participant's general familiarity with the product. In most cases, the participant should expect to spend approximately one hour reviewing the documentation for a walkthrough that will last 30-60 minutes. In addition to simply reading and reviewing the product, the participant should prepare and run a few test cases to see if it performs as it should.

Of course, a participant may feel somewhat unfamiliar with the product being reviewed or he may have questions about details of the program, the design, or the specification. In such a situation, the participant should plan on spending as much time as necessary *before* the walkthrough with the producer, and the producer should be prepared to devote as much time as necessary to familiarizing the participants with his product; this may require the producer to supply ancillary notes and documentation in addition to the product itself.

How can the participant be sure that he has devoted enough time to preparation for a walkthrough? How can the coordinator, and, most of all, the producer, be assured that the walkthrough will not degenerate into a superficial rubber-stamp approval of the product? One useful rule is that each participant should bring to the walkthrough at least one negative comment and one positive comment about the product being reviewed. Though this rule may seem simple-minded, my experience has been that it works: Each participant feels compelled— through peer pressure, if nothing else— to make at least one or two intelligent comments about the product. It is usually easy to think of something negative about another person's work, but more time and effort may be required to think of a positive, complimentary comment!

Another approach is to require the reviewers to prepare written notes. This can save a lot of time in the walkthrough by avoiding the need to discuss and review trivial errors and minor style suggestions. It also helps the reviewers avoid the sudden amnesia they face in the walkthrough when trying to remember the details of an error they noted when reviewing the product a day or two earlier.

QUESTIONS FOR REVIEW AND DISCUSSION

1. Why is it important that preparations be made before the walkthrough?

2. How many days in advance do *you* think a walkthrough should be scheduled? Do you agree with the suggestion of two days? What are your reasons?

3. What do you think the producer should be doing during the two days before a walkthrough? In your organization, is it easy to find other things to do while you are waiting?

4. What is the maximum-size product that can be adequately reviewed in 30-60 minutes? How many lines of code? How much of a structure chart? How many pages of dataflow diagrams?

5. What are the responsibilities of the coordinator before the walkthrough?

6. How can the coordinator ensure that the participants will invest the necessary time and energy to study the documentation before the walkthrough takes place?

7. What should be done if an invited participant fails to attend a walkthrough without a reasonable explanation? Should he not be invited to subsequent walkthroughs? Is this something that the project team should decide for themselves, or is it something that should be left up to management?

8. How much time do *you* think needs to be spent before the walkthrough in order to properly acquaint yourself with a product that is to be reviewed?

6 ACTIVITIES DURING THE WALKTHROUGH

There are moments when everything goes well; don't be
frightened, it won't last.

Jules Renard, *Journal*

Finally, we come to the walkthrough itself. The long
introduction is necessary and parallels the long buildup before
a *real* walkthrough. The objective, as I have pointed out, is to
provide as much advance preparation as possible, so that the
walkthrough will proceed quickly and efficiently.

It is important that the preparation not be wasted once
the walkthrough begins. The discussion and activities in the
walkthrough proper should be organized and methodical
(hence the name *structured* walkthrough) so that it does not
degenerate into an aimless argument between two
participants.

Two rules must be observed during the walkthrough for
it to succeed. First, everyone must agree to follow the same
set of rules and procedures, just as a convention or congress
agrees to follow Robert's Rules of Order. Second, the
participants must agree and understand that it is the *product*
that is being reviewed, not the *producer*.

I will discuss the major activities of the walkthrough in
terms of the primary participants, and then offer some
additional guidelines for successful walkthroughs.

6.1 Roles and Responsibilities in the Walkthrough

The coordinator begins the walkthrough by calling the assembled group to order. Though it is usually unnecessary, he may wish to remind the group of the nature and purpose of the walkthrough. In an organization that is carrying out walkthroughs of a number of different products, it is not unusual to see someone accidentally show up at the wrong meeting! More importantly, the coordinator may want to remind the participants that they have agreed to follow the same set of procedures during the meeting, and that they have agreed that it is the product, not the producer, being reviewed.

The presenter (who, as I pointed out in Chapter 4, is usually the producer) then takes the floor. If this is the first walkthrough of his product, he begins with a light overview of the program or design. Following that, he presents the product—whether it is a program listing, a structure chart, or a dataflow diagram—piece by piece, making sure that every piece is reviewed.

From the discussion in previous chapters, you can see that there may be several variations on this approach. For example:

1. If the reviewers have had ample opportunity to study the documentation before the walkthrough, then the producer's initial presentation can be brief—indeed, it may not even be necessary.

2. The reviewers may be concerned that the producer will brainwash them into making the same assumptions about the product that he did; consequently, they may prefer that he make no initial presentation. This approach also is taken in organizations where the reviewers wish to ensure that the product is "self-documenting," for the sake of maintenance.

3. If he is unsure, the producer may ask the reviewers whether or not they consider it necessary for him to provide an overview of his product.

4. In some cases, the producer may want to enlighten the reviewers about the background of the product, including alternative approaches that were considered, tradeoffs that were taken into account, and assumptions that were made. One could argue that this information should be included in the documentation provided to the reviewers before the walkthrough, but it is unlikely that it will be.[1]

If this is the second or subsequent walkthrough of his product, the presenter usually begins by reviewing old business, usually a point-by-point discussion of the bugs, suggestions, and comments raised in the previous walkthrough. For example, in a code walkthrough, the presenter/producer might say, "You'll see as we walk through this program that I fixed the bug that Fred found in the FICA-TAX module; you'll also notice that I followed Alice's suggestion to rewrite the nested IF statements in the NET-PAY module. And I did some research into Charlie's suggestion that we make the ZARK routine recursive; unfortunately, it increases the run-time by a factor of ten, so even though recursion would have made the module simpler and more elegant, I had to abandon it and stick with the original approach." Both the producer and the reviewers must have a detailed list of the comments that were raised in the previous walkthrough; this list is provided by the scribe, whose responsibilities are described below.

---

[1] Many people feel that today's CASE tools are just the primitive beginnings of "IPSE" tools (Integrated Programming Support Environments), and that the IPSE tools of the mid- to late-1990s *will* capture all of the alternatives, assumptions, and tradeoffs that are usually explored and discussed during early design activities, but then thrown away. This kind of information can be an incredibly valuable treasure-trove to later generations of maintenance programmers. It is extremely difficult, when looking at a program written five or ten years earlier, to tell *why* it was written the way it was.

The reviewers have an obvious responsibility in the walkthrough: to make constructive criticisms, comments, and suggestions. To avoid wasting time, the reviewers should give the producer and the scribe a list of errors that require no explanation— e.g., balancing errors in a set of dataflow diagrams, or an obvious, straightforward logic flaw in a program listing. Each reviewer then presents his comments about the product, making sure that his comments are directed at the product, rather than the producer.

Naturally, the producer will be tempted to argue about some of the comments, or defend his product, or challenge some of the assumptions made by the reviewers. As much as possible, he should resist this temptation because it consumes time and it raises the possibility of unproductive, unpleasant ego confrontations. Thus, the producer should not respond to any of the reviewers' comments in the walkthrough, except to ask for clarifications if necessary; he should *not* argue about the validity of the reviewers' comments. After the walkthrough is over, the producer will have time to review the suggestions in a more dispassionate mood. If he then feels it necessary, he can find one of the reviewers for a private discussion or argument about comments raised in the walkthrough.

Throughout the walkthrough, the scribe records the reviewers' comments. Figure 6.1 shows a typical walkthrough summary, and Figure 6.2 shows a typical completed form. In addition, the scribe keeps a detailed "action list" of the errors and suggested changes to the product. However, for reasons discussed in more detail in Chapter 12 (the desire to prevent walkthroughs from playing an unwanted role in employee performance reviews), the action list should be kept separately from the summary; the summary is the public record of the walkthrough and is available for everyone to see.

A typical action list is shown in Figure 6.3. Note that the comments are recorded in a summary form; the scribe does not have the time for verbose comments, nor should the reviewers require such verbosity. Also, note that each action item has a "disposition code" indicating how the reviewers feel the error should be dealt with. Some examples of typical disposition codes are as follows:

S   *Suggestion.* The producer may accept the suggestion or decide to ignore it at his pleasure. Most issues of style would have this disposition code.

D   *Defect.* This is an error that the reviewers feel *must* be corrected, and that cannot be ignored by the producer. Violations of a "hard" standard also fall into this category. Issues of style would normally not have this disposition code unless the walkthrough participants felt strongly enough to take a vote. (As discussed below, the group must decide in advance whether a majority vote, unanimous vote, etc., is required in cases like this.)

O   *Open issue.* Some issues that are raised in the walkthrough are not black and white, and cannot be given a "D" code; on the other hand, they are too important to be dismissed with an "S" code. An example of this is a suggestion by one of the walkthrough participants that the producer could use a different algorithm in his program that *might* improve the speed by two orders of magnitude. If the suggestion can't be resolved within a few minutes, and if the reviewers feel it could have a dramatic effect on the final product, then it becomes an "O." Action items of this kind virtually guarantee that the product will require a subsequent walkthrough, in order to see how the producer has resolved the open issue.

When the walkthrough has finished (that is, each of the reviewers has had a chance to make comments), the coordinator should ask the group members for their recommenda-

```
┌──────────────────────────────────────────────────────────────┐
│ Coordinator                                                    │
├──────────────────────────────────────────────────────────────┤
│ Project                                                        │
├──────────────────────────────────────────────────────────────┤
│ Coordinator's checklist:                                       │
│ 1.    Confirm with producer(s) that material is ready and      │
│       stable:_____                                       │
│                                                                │
│ 2.    Issue invitations, assign responsibilities,             │
│       distribute materials                                     │
│                                                                │
│       Date:_____    Time:_____    Duration:_____    │
│       Place:_____  │
│                                                                │
│       Responsibilities  Participants                           │
│                                         Can     Received       │
│                                         attend? materials?     │
│                                                                │
│       1_____      _____  ☐         ☐         │
│       2_____      _____  ☐         ☐         │
│       3_____      _____  ☐         ☐         │
│       4_____      _____  ☐         ☐         │
│       5_____      _____  ☐         ☐         │
│       6_____      _____  ☐         ☐         │
├──────────────────────────────────────────────────────────────┤
│ Agenda:                                                        │
│                                                                │
│ _____1.    All participants agree to follow the same set of  │
│              rules.                                            │
│ _____2.    New project: walkthrough of material              │
│              Old project: item-by-item checkoff of previous    │
│              action list                                       │
│ _____3.    Creation of new action list (contributions by each│
│              participant)                                      │
│ _____4.    Group decision                                    │
│ _____5.    Deliver copy of this form to project management   │
├──────────────────────────────────────────────────────────────┤
│ Decision:     _____Accept product as-is                      │
│               _____Revise (no further walkthrough)           │
│               _____Revise and schedule another walkthrough   │
├──────────────────────────────────────────────────────────────┤
│ Signatures:            │                    │                  │
│                        │                    │                  │
│                        │                    │                  │
└────────────────────────┴────────────────────┴──────────────────┘
```

Figure 6.1: **A typical walkthrough report form**

Coordinator	*Marcia Greenberg*

Project	*Zarcon Cash Flow Analysis System: George's Fabulous Lookup Subroutine*

Coordinator's checklist:
1. Confirm with producer(s) that material is ready and stable ✓

2. Issue invitations, assign responsibilities, distribute materials
    Date: *Jan 29*   Time: *10:00*   Duration: *30 min*
    Place: *Conference room next to the room where they make popcorn*

Responsibilities	Participants	Can attend?	Received materials?
1 *Coordinator*	*Marcia*	✓	✓
2 *Presenter*	*Pete*	✓	✓
3 *Scribe*	*Sue (under protest!)*	✓	✓
4 *Maint. oracle*	*Johann*	✓	✓
5 *Producer*	*George*	☐	☐
6_____	_____	☐	☐

Agenda:

✓ 1. All participants agree to follow the same set of rules.
✓ 2. New project: walkthrough of material
       Old project: item-by-item checkoff of previous action list
✓ 3. Creation of new action list (contributions by each participant)
✓ 4. Group decision
✓ 5. Deliver copy of this form to project management

Decision:	__ Accept product as-is
	✓ Revise (no further walkthrough)
	__ Revise and schedule another walkthrough

Signatures:	*Marcia Greenberg*	Pete Peterson
Sue Henderson	Johann Spritzler	George

Figure 6.2: **A typical completed walkthrough form**

Coordinator	Producer	
Project		
Date	Time	

ACTION   ITEMS

#	Disposition	Description
1.	☐	_____
2.	☐	_____
3.	☐	_____
4.	☐	_____
5.	☐	_____
6.	☐	_____
7.	☐	_____
8.	☐	_____
9.	☐	_____

Figure 6.3: **A typical action list**

tions concerning the product. Normally, four recommendations are possible:

- The group may vote to accept the product in its present form.

- The group may vote to accept the product with the revisions and corrections suggested in the walkthrough, implicitly trusting the producer to make the changes correctly.[2]

---

[2] However, keep in mind that studies of software maintenance indicate that, at best, there is only a 50 percent chance that a programmer will be successful *on his first attempt* to correct an error in his program. Reviewing the changes to a program or any other product associated with a systems development project is no different, and implies no more criticism of the producer than proofreading the corrections to any other written document.

- The group may vote that another walkthrough is necessary, either because a substantial number of errors were found and the group is concerned that their correction may introduce new errors, or because some controversial suggestions and criticisms (of either the style or the content of the product) were raised.

- My colleague Nan Matzke points out that in rare cases, the product may be so bad that the reviewers will vote that it be thrown away and rewritten. This could also happen if the reviewers had an initially positive reaction toward the product but found that the error density was unacceptably high— e.g., 27 errors in a module consisting of 13 COBOL statements.

Whether the vote should be unanimous, a simple majority vote, or a non-enforceable record of the reviewers' opinions depends on the nature of the organization. In a programming team, the entire team assumes responsibility for the correctness and quality of work of any of its members, so the vote to approve a product would have to be unanimous. In a more loosely formed "semi-team," a majority vote might be sufficient. In an organization where the producer assumes total responsibility for his work, the reviewers should merely offer their opinions, leaving the producer to decide whether or not additional walkthroughs are necessary.[3]

6.2   Additional Guidelines

By now, you should be able to conduct a successful walkthrough; however, that doesn't mean that it will be trouble-free. In subsequent chapters, I will discuss personality and psychological problems which can ruin an otherwise successful walkthrough. For the remainder of this

---

[3] Even in the last case, there is a subtle pressure on the producer to accept the suggestions of the reviewers. Think about it: If you attend two consecutive walkthroughs of Fred's coding efforts and find that Fred categorically refuses to consider your suggestions, why would you bother attending any subsequent walkthroughs? For that matter, why would he invite you?

chapter, I will discuss guidelines that can help avoid the more mundane problems encountered in a walkthrough.

1. _Keep the walkthrough short._ Walkthroughs are mentally fatiguing. Reviewers cannot expect to maintain their concentration for more than an hour or two hours at the very most. Try to keep the walkthrough to 30 minutes if possible; indeed, there is nothing wrong with a 15-minute walkthrough. Among other things, you will find that the reviewers will be more willing to participate if they know it will be brief; also, the coordinator will find it easier to arrange logistics for a short walkthrough. But the most important benefit is the increased concentration, which almost certainly leads to a better walkthrough than one where eyelids are beginning to droop.

2. _Don't schedule more than two consecutive back to back walkthroughs._ This suggestion follows from the previous comments. One can expect the first walkthrough to be very productive, the second acceptable. By the third walkthrough, the reviewers are mentally exhausted, and by the fourth, they are sound asleep (mentally, if not physically). Even if the walkthroughs are as short as 15 minutes, mentally switching gears from one topic to another makes it difficult to have more than two consecutive walkthroughs. A one-hour break usually is sufficient for everyone to revive and prepare for another round.

3. _Don't walk through "fragments" of a product._ Though this suggestion should be obvious, I have personally watched MIS professionals attempt to walk through a 50-statement fragment of an unfinished program, or the first three pages of a detailed narrative specification of a system. In most cases, this occurs because the producer felt obliged to meet a walkthrough deadline, regardless of whether he was ready for it. From the reviewers' point of view, the walkthrough of a fragment serves

very little purpose and is enormously frustrating. The moral: Review *complete* pieces of a product, such as the code or design for an entire module or for a meaningful subsystem that can be viewed in isolation from the rest of the system.

4. *Use standards to avoid disagreements over style.* Despite the admonitions earlier in this chapter, it is common for arguments to break out between the reviewers, or, more likely, between the reviewers and the producer. Sometimes the argument will revolve around a bug— e.g., whether or not the program is really operating correctly, or whether the systems specification reflects the user's true requirements. However, many of the arguments concern style. Reviewers may disagree about the maximum number of bubbles that are appropriate for a dataflow diagram, or they may argue about the suitability of a certain construct in COBOL or Ada, or they may criticize the graphic conventions used in a structure chart ... or they may argue about whether it is important to rewrite a module to save 128 bytes of memory and 30 microseconds of CPU time. The producer has two problems with style disagreements: First, he may sincerely feel that his style is appropriate, particularly if he has considered several alternatives privately before coming to the walkthrough; second, the psychological and emotional commitment to his product makes it difficult to scrap it and begin anew.

Though I will return to the psychological problems in later chapters, there is a procedural approach that can help avoid the nastier conflicts about style and format: standards. The very word "standards" sends a shiver up the spine of the average computer professional, for he feels that a 17-volume tome on standards has no relevance to his day-to-day work. Thus, it is ironic that several organizations have begun to see standards in a whole new light— as a way of avoiding fistfights in a walkthrough.

In most cases, the reviewers do not need to use the entire 17 volumes of "official" standards; it is far more common for the group to extract the few pages that apply to their product or to make their own ad hoc style guide. The informal standards are usually only a few pages long, and they can be developed using the same walkthrough approach as for other products.

Of course, it would be time-consuming and expensive to develop a set of standards for each new product being reviewed. However, it makes a great deal of sense to develop a common set of standards for all of the systems analysts, designers, and programmers working on a single, common MIS development project.

5. *Agree ahead of time to respect the coordinator's role in the walkthrough.* When a group of intelligent, articulate, aggressive computer people gets together for a walkthrough, it is easy for things to get out of hand. MIS people just seem to have a natural talent for arguing! In the midst of their arguments, they may be reluctant to heed the coordinator's plea to maintain order and decorum, especially if the coordinator is a peer without the authority to order the participants to shut up. The only way the group can avoid the risk of turning the walkthrough into a noisy brawl is to agree ahead of time that they will respect the coordinator's role and that they will stop arguing if the coordinator asks them to. If this cannot be done, one of two things will eventually happen: (a) the walkthrough will be perceived by everyone as unproductive and will be abandoned, or (b) management will insist that a project leader who *can* order the participants to shut up play the role of coordinator. As we will see in Chapter 11, there are many reasons why it is preferable *not* to have the project leader present; it is well worthwhile for the project team to learn to police itself.

6. _Remember that the basic purpose of the walkthrough is error DETECTION, not error CORRECTION._ In some cases, the mere description of what the error is will be sufficient for the author to understand how to fix it; indeed, some errors are so trivial that they can be written in advance by the reviewers and submitted to the scribe at the beginning of the walkthrough without any discussion at all. In other cases, the existence of the error can be easily stated, but its correction may not be so obvious. A little creative brainstorming in the walkthrough may be productive, but it usually degenerates very quickly: There is hardly anything less productive than watching six programmers simultaneously dashing up to a blackboard to begin scribbling their proposed "fix" to a bug.

QUESTIONS FOR REVIEW AND DISCUSSION

1. What are the two most important rules for ensuring the success of a walkthrough?

2. Do you think that the producer should begin a walkthrough by presenting his product to the reviewers? What are the advantages and disadvantages?

3. Are there any situations where the producer definitely should NOT present his product to the reviewers?

4. Why should the producer not argue about the criticisms and suggestions raised by the reviewers?

5. What recommendations can the reviewers make at the end of the walkthrough? Can you think of any in addition to the three mentioned in this chapter?

6. Should the reviewers' recommendations be based on a unanimous vote, or should they be the result of a simple majority vote?

7. Why should walkthroughs be kept to 30-60 minutes? Is there anything wrong with a walkthrough that is significantly shorter than 30 minutes—e.g., 10-15 minutes?

8. How many consecutive walkthroughs do you think can be held before the quality of the walkthroughs deteriorates?

9. How can standards be used to avoid arguments over the style of the product? Are there any other ways to avoid such arguments?

7 ACTIVITIES AFTER THE WALKTHROUGH

It is said an Eastern monarch once charged his wise men to invent him a sentence to be ever in view, and which should be true and appropriate in all times and situations. They presented him with the words: "And this, too, shall pass away." How much it expresses! How chastening in the hour of pride! How consoling in the depths of affliction!

Abraham Lincoln
Address to the Wisconsin State Agricultural Society,
Milwaukee, September 30, 1859

This chapter will be very brief. Indeed, you might wonder why it exists at all. What is there to say about a walkthrough *after* the walkthrough is over?

That's just the point. A lot of people think that nothing remains to be done once the walkthrough is finished. As a result, small, but nevertheless important, loose ends are not taken care of. Hence this chapter, to emphasize those few details that need to be done to properly complete the walkthrough.

As in previous chapters, I will discuss these post-walkthrough activities by outlining the responsibilities of the various participants. As with the activities in Chapter 5, these post-walkthrough activities are primarily administrative and clerical; thus, they could be accomplished easily and efficiently with appropriate office automation facilities. In today's typical MIS organization, however, a certain amount of paperwork and manual effort will still be required.

7.1   Responsibilities of the Coordinator

The coordinator has the important post "clean-up" activities. He or she divides the walkthrough report that we saw in Figure 6.1 into two distinct pieces: a management summary and a set of detailed comments made by the participants.

The management summary, shown in Figure 7.1, normally is delivered to the project manager or to the person who is responsible for monitoring progress of the product under review.[1]   Note that the management summary does *not* indicate how many errors were found, nor does it reveal the detailed comments and criticisms made by the reviewers. Instead, it simply states what was reviewed, when the walkthrough took place, who attended, and what the verdict was.   This information is usually more than sufficient to give management an adequate feeling for the status of the product.

As for the detailed comments made by the reviewers, it is generally agreed that management wouldn't understand them and wouldn't be interested in them.   I feel that management should not see the detailed comments because once it becomes known that they are looking, the objectivity of the walkthrough is severely threatened.   I will have much more to say about this phenomenon in later chapters.

The detailed comments should be saved in case they are needed in the future.   Most organizations feel that the detailed comments, *plus* a copy of the management summary, should be filed along with the "official" copy of the documentation itself—that is, along with the official program listing, the structure charts, or the dataflow diagrams.   In many projects, this documentation would be maintained by a project librarian, so the coordinator would simply turn over the detailed notes to the librarian; in the best of cases, the comments would be stored in electronic form in an on-line project library.

---

[1] Figure 7.1 is the same as Figures 6.1 and 6.2.

```
Coordinator    Marcia Greenberg

Project        Zarcon Cash Flow Analysis System: George's Fabulous
               Lookup Subroutine

Coordinator's checklist:
1.      Confirm with producer(s) that material is ready and
        stable  ✓

2.      Issue invitations, assign responsibilities, distribute
        materials
        Date: Jan 29        Time:  10:00       Duration:  30 min
        Place: Conference room next to the room where they make popcorn

        Responsibilities        Participants        Can      Received
                                                    attend?  materials?

        1 Coordinator           Marcia              ✓        ✓
        2 Presenter             Pete                ✓        ✓
        3 Scribe                Sue (under protest!) ✓       ✓
        4 Maint. oracle         Johann              ✓        ✓
        5 Producer              George              ☐        ☐
        6                       _____             ☐        ☐

Agenda:

✓    1.    All participants agree to follow the same set of
           rules.
✓    2.    New project: walkthrough of material
           Old project: item-by-item checkoff of previous
           action list
✓    3.    Creation of new action list (contributions by each
           participant)
✓    4.    Group decision
✓    5.    Deliver copy of this form to project management

Decision:      __   Accept product as-is
               ✓    Revise (no further walkthrough)
               __   Revise and schedule another walkthrough
Signatures:              Marcia Greenberg          Pete Peterson
Sue Henderson            Johann Spritzler          George
```

Figure 7.1: **Management summary of a walkthrough report**

Finally, the coordinator is responsible for delivering a copy of the detailed comments to each of the participants. The participants should have their copies as quickly as possible, preferably within an hour after the walkthrough is over, so they can review them while the details are still fresh in their minds; of course these copies could also be delivered via electronic mail if the MIS organization has the appropriate technology.

Once all of this has been done, the coordinator can turn himself into a toad and vanish.

7.2   Responsibilities of the Participants

The responsibilities of the participants depend largely on the producer. In most cases, the participants have nothing to do—beyond reviewing the copy of the detailed comments they receive from the coordinator to ensure that there are no errors.

However, occasionally the producer wishes to discuss some detailed points that were not fully resolved in the walkthrough itself.   If, for example, the participant commented in the walkthrough that he felt that 16 levels of nested IF statements were difficult to understand, he might find that the producer wanted some help in finding an alternate, more reasonable, approach. And if the participant found a subtle bug in the walkthrough, there's a good chance that a *solution* to the bug was not offered in the walkthrough (most likely because a solution was not immediately obvious, and everyone felt it would be unproductive to try to find one at the time). Thus, the participant may find that the producer approaches him after the walkthrough, asking for help in fixing the bug.

While technically the participant has no obligation to help the producer solve his problems, I feel that he has a moral obligation to spend up to an hour explaining the comments and criticism that he made in the walkthrough, particularly if the participant's original comments were in the form of "I disagree with the style of this product because..." or "My opinion is..."

Possibly the producer may try to get the participant to solve *all* of the problems raised in the walkthrough. It is important that the participant remember that the producer is the one who "owns" the problem, unless the product is being developed by a true programming team, in which case the members all own the problem. In a conventional MIS organization, though, it is perfectly appropriate for the participant to say, "Hey, Charlie Producer, I'd be happy to brainstorm with you for half an hour, but I can't spend much more time than that: I've got a lot of other work I've got to finish this afternoon. I'm afraid you're gonna have to go off in a corner and figure this out for yourself."

7.3   Responsibilities of the Producer

Finally, the producer: What responsibilities does he have? From the comments above, we can see that his primary responsibility is to ensure that he understands the comments, criticisms, and suggestions that were made during the walkthrough and that have been documented in the walkthrough report. He doesn't have to agree with all of the comments— and he doesn't have to have instant solutions for all of the problems that were cited. But he should at least understand what everyone was saying. If he doesn't, he should have the opportunity— and the responsibility— to pester the participants until he *does* understand.

What then? Assuming that the producer probably knows more about his product than the participants, he's the one who will have to decide which suggestions can be accommodated easily and which ones are impractical. Thus, the producer should consider seriously and objectively each suggestion, each criticism, and each complaint and accept those which seem appropriate, compromise where it seems reasonable, and reject those which turn out to be unreasonable. Of course, if his product was rejected by the reviewers, he has to take into account that the same suggestions and criticisms may be raised in a subsequent walkthrough; consequently, the producer may decide to accept some suggestions even though he doesn't completely agree with them. If he strongly disagrees, he has to devise a rational argument that he can present in the next walkthrough.

The final step for the producer, assuming that the product was not completely accepted, is to start the cycle over again; that is, he has to decide when he's ready for another walkthrough. As I have pointed out, the next walkthrough typically will begin with a review of old business, so the producer needs to be able to show— either in the product itself, in the accompanying documentation, or in his oral presentation— that the bugs have been fixed and that the suggestions for improvement have been seriously considered.

One last note: The producer should personally thank the coordinator, the scribe, and the other attendees for participating in the walkthrough; after all, they have devoted a significant amount of their time to helping him improve the quality of his work.

QUESTIONS FOR REVIEW AND DISCUSSION

1. What responsibilities does the coordinator have when the walkthrough is finished?

2. What responsibilities does the producer have when the walkthrough is over?

3. Why should the detailed comments be distributed to the participants as quickly as possible after the walkthrough is over?

4. How much time do you think the participants should be willing to spend helping the producer resolve the problems that were brought up in the walkthrough?

8 OBSERVATIONS ABOUT PROGRAMMERS AND ANALYSTS

Previous chapters have discussed the *concepts* and *mechanics* of walkthroughs: when they are scheduled, how they are organized, and who does what to whom. If every programmer and systems analyst is familiar with the rules, the roles, and the procedures, all walkthroughs should be smashing successes, right?

Unfortunately not. There is more to a walkthrough than rules and procedures. A typical walkthrough involves a variety of psychological issues, many of which are only vaguely understood by the participants. This chapter and the next are concerned with these psychological issues.

Before we begin, let's put things in the proper perspective. I am neither a psychologist, a psychiatrist, nor a psychoanalyst, so any comments that I make in this chapter must be interpreted as those of an amateur observer of people working in the data processing field.[1] However, I assume that you are not a psychologist or psychoanalyst either; so even if you got advice from a professional psychiatrist, you would have to interpret it in your own amateurish way.

Though we may be laymen psychiatrists, we all have substantial experience in the data processing field, and we have met a variety of programmers and systems analysts. By remembering what kinds of personalities you have run into in your career, you can be prepared for many of the personality clashes that might otherwise ruin a walkthrough.

8.1   What Are Programmers and Systems Analysts Really Like?

Several years ago, I worked on a systems development project with a group of programmers and systems analysts whose personalities were almost impossible to cope with. As a group, they tended to be arrogant, opinionated, insecure, and generally anti-social.[2]   What made them behave the way they did? The only explanation that made sense to me was that each of them had been attracted to the computer field because he preferred dealing with machines rather than people. Computers are cold, rational, deterministic objects that don't talk back, whereas people are irrational, emotional, unpredictable organisms that *do* talk back.

You've probably met a few arrogant, opinionated, insecure, anti-social programmers and systems analysts, too. But it would be foolish to conclude that they are *all* arrogant so-and-so's. In fact, it's dangerous to make any generalizations about the personalities of programmers and systems analysts because there are now so many people in the

---

[1] For more expert opinions on the attitudes and personalities of computer programmers, as well as "ordinary" people who come into contact with computers, see [Brooks, 1975], [Weinberg, 1971], [Turkle, 1984], [Thomsett, 1981], [Cougar and Zawacki, 1980], and [Semprevivo, 1981].

[2] And of course I regarded myself as a humble but competent programmer who could get along with *anyone*, as long as that "anyone" was reasonable. It's virtually impossible to be unbiased when making observations about the personality of one's peers!

computer field, from so many walks of life, that one can find examples of almost any kind of personality.

Many programmers and systems analysts are nice, responsive, cooperative, friendly, intelligent, and articulate— as well as loyal, thrifty, and brave. But they are not the ones who are going to cause trouble for you in a walkthrough. On the contrary, they are the ones who will make a walkthrough useful and fun.

It's those other people we're concerned about. Some of their more important personality traits are discussed in this chapter.

8.2   Programmers and Systems Analysts Have High IQs

This statement is obviously a generalization; some incredibly stupid people have managed to get jobs in the computer industry.[3] But it is common to see at least one or two people in an MIS organization who are— or at least seem to be— geniuses. They have keen, analytical minds; they can carry out calculations in their heads almost as quickly as the computer can; and they seem to have photographic memories.

A genius can be an enormous asset in a walkthrough. He can spot bugs that others wouldn't find, suggest better ways of solving the user's problem, more elegant ways of designing or coding a system, and so forth. But if his intelligence is combined with a few other personality traits that will be discussed below (such as impatience and arrogance), the genius can be highly destructive. He can easily humiliate or insult the producer and the walkthrough participants, and, in

---

[3] There will probably be fewer stupid people in the computer field as time passes, that is, assuming that the current generation of computer programmers and systems analysts (many of whom have no education, or a graduate degree in basket weaving or ancient Sanskrit) retire and fade into oblivion. The education of today's computer professionals in the university, and even in the secondary schools, is vastly superior to what it was 20 years ago, and the enormous popularity of the computer field has made it possible for most universities to be selective about the people they admit into their Computer Science programs. On a broader scale, it's worth noting that long division was a college subject in medieval times; now it's taught in elementary school. Similarly, structured design was a graduate level subject only a few years ago; now high school students learn the concepts in many cities around the U.S.

the extreme, his verbal barbs can destroy the producer's sense of self-esteem.

Also, often his expertise is restricted to a very narrow field. Just as the idiot savant can multiply ten-digit numbers in his head and yet not be able to read, write, or spell, so the genius programmer may be able to memorize 500-page hexadecimal dumps without being able to express his thoughts in COBOL or Pascal or English. This narrowness is frequently encouraged in MIS organizations. We find database experts—or more narrowly still, DB2 experts, ADABAS experts, or even experts in a specific "release" of IMS or TOTAL. Similarly, there are operating system gurus and teleprocessing wizards who know almost nothing outside their narrow specialty. Conversely, we find users who are experts in their fields but who miss the possibilities computers can offer them.

The "specialist" genius can create havoc in a walkthrough, for his comments and criticism often fail to take the big picture into account. For example, the operating system expert may argue that the producer's design is wasting a substantial amount of CPU time—without appreciating that CPU time is relatively cheap and that various other factors like development time and maintainability may be considerably more important.

How can you deal with the genius in a walkthrough? Deference and humility are probably the two best attitudes to assume. After all, if Suzanne is ten times smarter than you—if only in a single, narrow subject area—you might as well accept her criticisms and suggestions gracefully. In addition, there's no point in arguing about an issue unless you're absolutely sure of yourself, and even if you're right, the genius can often dazzle you with verbal pyrotechnics.

Remember also that you're not supposed to argue about things in a walkthrough, regardless of whether the suggestions are made by a genius or a moron. The primary purpose of the walkthrough is to *raise* issues, not to resolve them on the spot. So let the genius make suggestions and criticisms, without any attempt to understand them or rebut them; at your leisure, you can plod through her reasoning and decide whether she's right.

8.3   Programmers and Systems Analysts Are Impatient, Arrogant, and Opinionated

I could discuss each of these personality traits separately, but they are so intertwined that one often can't distinguish arrogance from impatience. Is it impatience, for example, that makes a systems analyst interrupt one of his fellow participants in mid-sentence to make a comment about the product being reviewed? Or is it arrogance? Is it arrogance that makes a programmer try to cut off discussion about some issue by saying, "Well, it's *intuitively* obvious that..."? Or is it just a sign that he is opinionated?

As I have mentioned, these personality traits are often combined with high intelligence. The genius becomes impatient when his colleagues— mere mortals that they are— are slow to understand a point he is making. He is also likely to make comments that will be interpreted by his colleagues as arrogant: "Well, if you knew anything about the Pascal language, you would *know* that this statement is incredibly inefficient. Only a bozo would write such a statement!" Further, the genius may have a highly developed intuition, a sixth sense, that enables him to spot bugs or find better solutions to a problem— but since he can't *prove* that his intuition is right (or is unable to articulate a proof), he appears to his colleagues to be highly opinionated.

How does one deal with the arrogant, opinionated, impatient systems analyst? In the walkthrough itself, it's probably best not to respond to inflammatory remarks that such a person may make; let the scribe record his opinions, regardless of the arrogance with which they were expressed, and don't let an argument begin. This restraint becomes a major responsibility of the coordinator, of course, who also has to ensure that the reviewer's impatience doesn't disrupt the agenda of the walkthrough.

Outside the walkthrough, the participants should try to explain to the programmer/analyst how they feel about his arrogance and impatience. If the participants are working together as a programming team, then they have a responsibility to resolve their personality clash. But it's not the sort of problem that can be resolved in the middle of reviewing a product in a walkthrough, nor is it the sort of thing that is likely to be resolved in one confrontation. It may be helpful to videotape one of the walkthroughs, and then let all of the participants review the videotape so that the personal interactions can be reviewed and discussed; often, people are largely unaware of the impact they have on others.

If the group is *not* working as a true programming team, then there are several options if the personality clash cannot be resolved. One of the most effective ones is for the producer to avoid inviting the arrogant programmer/analyst to any future walkthroughs!

8.4 Programmers and Systems Analysts Are Insecure and Defensive

In some cases, the arrogant, opinionated programmer/analyst merely is covering up his insecurity. If he can't defend his point with rational logic, he blusters and tries to win the argument by insulting his opponent. If, as a producer, he sees that another participant is about to expose a major flaw in his product, he shows impatience and does everything he can to curtail discussion to minimize his embarrassment.

Note that it is the producer who is most likely to show signs of insecurity and defensiveness in a walkthrough because it is his product that is being exposed to public scrutiny. The other participants can cover up any insecurity they feel by remaining silent in the walkthrough; if they don't spend much time criticizing the producer's work, nobody will realize how unsure they feel about the field in which they work. But the producer has no such opportunity, and it is normal for a person in *any* profession to be insecure occasionally about the quality of his work.

One characteristic of the computer environment that emphasizes the normal human feelings of insecurity is the speed with which today's technology becomes obsolete. A person who was both competent and confident in the 1970s finds that he has to deal with a bewildering array of new technologies. Thus, if he proposes a technical solution to a user's information processing problem, he always has to wonder if a better solution was published in last week's *Computerworld*. And as he looks forward to the 1990s, he has to worry about a whole new generation of programmers and systems analysts who may be better trained and more comfortable than he is in artificial intelligence, relational database technology, and a dozen other technical specialties.

How can you deal with the insecure, defensive programmer or systems analyst? Since he's likely to be the producer, remember that the walkthrough is supposed to be conducted in a non-threatening, non-embarrassing fashion. So if criticisms are raised, they should be raised in the form of suggestions that the producer can think about in the (psychological) safety of his own office. Rather than saying, "I think that this program is pathetically inefficient, and I wouldn't accept it," a reviewer might say, "I think this program might run ten times faster if you used the Widget Algorithm, but I'm not sure I understand the complexities of your program well enough to know whether the Widget Algorithm would work. Perhaps you could look into it; there's a really great description of the algorithm in last month's *IEEE Transactions on Software Engineering*."

If the programmer/analyst becomes defensive about a criticism or suggestion and begins to argue, remember that arguments should be avoided in walkthroughs. The coordinator should watch for such arguments and stop them immediately.

8.5 Programmers and Systems Analysts Are Conservative and Tradition-Bound

As I noted, some programmers and systems analysts show signs of being insecure about their work; as a result, they sometimes become defensive when presenting their product in a walkthrough. Another common consequence of this insecurity is a strong tendency to follow a traditional approach to solving a problem. Thus, if the programmer or systems analyst did his early work in the 1970s designing business data processing systems with IBM 360-style sequential file systems, then he may continue following the same approach on a vintage-1988 IBM 3090 with DB2. Similarly, if a programmer found that his COBOL compiler compiled nested IF statements incorrectly five years ago, he might continue to reject techniques like structured programming for the next ten years.

This reaction may become apparent if the programmer or systems analyst is a presenter in a walkthrough: The other participants will notice that he has followed an antiquated approach to solving the problem, and that he defends it vigorously. Of course, if the tradition-bound programmer is reviewing someone else's work, he might say, "That approach won't work here. We tried it a few years ago, and it was a disaster."

How should this response be handled in a walkthrough? Many of the comments made above apply in this case: Everyone, especially the coordinator, should ensure that the tradition-bound programmer does not start an argument about the "good old days" and the virtues of the IBM 1401. At the same time, it is usually a good idea to listen to what the traditionalist is saying. We *do* have a tendency in the data processing field to become fascinated with gadgets, fads, and buzzwords, and it is often useful to have someone remind us of the fundamental principles upon which all successful MIS systems are built.

8.6   Programmers and Systems Analysts Are Uninterested in the "Real World"

As you can imagine, this criticism usually is aimed at programmers rather than systems analysts; after all, the primary responsibility of a systems analyst is to describe the real-world problems that a user faces. A programmer or systems designer, on the other hand, can insulate himself from the user and other representatives of the real world and spend his time tinkering with hardware and software.

Indeed, "tinkering" is an apt word. Many programmers play with computers as if they were toys and approach computer programs as if they were a form of crossword puzzle.   In companies where I have worked or consulted, these people were often referred to as hackers, "tech weenies," or code bums, and while it may have been an unpleasant description, it fit.

The hacker can waste time in a walkthrough arguing about esoteric features of the product which have nothing to do with the real-world requirements.   Since almost everyone has a little of the hacker personality in him, it is easy for the entire group to begin arguing about such things as the possibility of using recursive subroutines to solve the user's problem.   While this is interesting, it is usually irrelevant.

8.7   Summary

As we have seen in previous chapters, walkthroughs have procedures and standards; and the products being reviewed are dry, technical documents. But you can't remove the human element from a walkthrough. The producer, whose "work of art" is being reviewed, has human feelings and emotions.   And the reviewers have feelings of their own.

This chapter has begun to show you why a walkthrough might not be a calm, rational, objective discussion of the technical merits of a systems development product. Chapter 9 continues this discussion by showing you some of the common psychological games that take place in a typical walkthrough.

QUESTIONS FOR REVIEW AND DISCUSSION

1.   Do you think that personality conflicts play a major role in walkthroughs? Why?

2.   On the average, do you think that the programmers and systems analysts in your organization are easy to get along with, or do their personalities make communication difficult?

3. If there are personality clashes in your organization, do you think they can be resolved by the individuals themselves? Or do walkthroughs require a resident psychiatrist to resolve problems?

4. Would you classify any of the technicians in your organization as geniuses? What are the advantages and disadvantages of having a genius participate in a walkthrough?

5. Do you think that any of the programmers or systems analysts in your organization are arrogant, impatient, or opinionated? If so, how can you deal with them in a walkthrough?

6. Are any of the programmers or systems analysts in your organization insecure or defensive? How can you deal with them in a walkthrough?

7. Do any of the programmers or systems analysts in your organization have other personality traits that make walkthroughs difficult?

9 GAMES PROGRAMMERS PLAY

> The great wish of some is to avenge themselves on some particular enemy, the great wish of others is to save their own pocket. Slow in assembling, they devote a very small fraction of the time to the consideration of any public object, most of it to the prosecution of their own objects. Meanwhile each fancies that no harm will come of his neglect, that it is the business of somebody else to look after this or that for him; and so, by the same notion being entertained by all separately, the common cause imperceptibly decays.
>
> Thucydides
> *The History of the Peloponnesian War,* Book I, section 141
> 431-413 B.C.

In the preceding chapter, I made a number of general observations about the personalities of programmers and systems analysts, and we saw that some of these traits could have a serious impact on the effectiveness of walkthroughs. In this chapter, I will address this same issue more directly: I consider the *games* that programmers and systems analysts play when in a walkthrough, and in dealing with one another outside the walkthrough. Regardless of your position in an MIS organization—junior programmer, senior systems analyst, or manager—you should have some awareness of these psychological games, so that you can deal with them when you encounter them.

9.1   The Concept of Games

The concept of games between people was popularized by Eric Berne in *Games People Play* [Berne, 1964]. As Berne puts it, "A game is an ongoing series of complementary ulterior transactions progressing to a well-defined, predictable outcome." Or, more appropriately, a game is "a recurring set of transactions, often repetitious, superficially plausible, with a concealed motivation; or more colloquially, a series of moves, with a snare, or gimmick."

Games take place among people of all types, and they take place in every conceivable social situation. Berne identifies life games, marital games, party games, sexual games, underworld games, consulting room games, and even good games. Outside the data processing profession, all of us play games with our friends, our enemies, our spouse, our neighbors, and occasionally even strangers. But they take place inside the data processing profession, too, and that is the subject of our discussion in this chapter.

Games have been played between one programmer and another, between a programmer and his manager, and between various others long before Berne introduced the buzzword and categorized many of the more common social games. But as I observed in the previous chapter, many programmers and systems analysts are basically anti-social, and by avoiding one another whenever possible, they may have avoided participating in the more popular games (though, as Berne points out, this anti-social behavior is often manifested in a game called "See What You Made Me Do"). In any case, the introduction of walkthroughs and programming teams has forced many programmers and systems analysts to deal with one another on a level that is new to them. The games, which previously could be ignored, have become a major element in their day-to-day work.

One of the basic theses of *transactional analysis*— the analysis of interchanges which form games— is that all people operate on three different psychological levels: a Parent, an Adult, and a Child. At one time or another, for example, each of us functions as a parent— that is, we respond as we remember or perceive that our parents responded, and we frequently use the same posture, gestures, and vocabulary that they did. In other circumstances, each of us acts as a child: spontaneous, rebellious, dependent, creative, and intuitive by nature. On rare occasions, we function as adults: logical, rational, and unbiased in our analysis of the information given to us. Since this book is not a detailed treatise on psychology, we will not explore these concepts any more deeply. Suffice it to say that programmers and systems analysts exhibit Parent, Adult, and Child personalities as clearly as non-computer people do.

9.2   Typical Games

Berne identifies some 35 distinct games that take place among people; all of them can take place (and probably *do* take place) between programmers and systems analysts *outside* the office and the computer room. Many of these classical games take place *in* the office, too, and they may take place in a walkthrough.

Since this book is about walkthroughs, it might seem more appropriate to discuss the games that actually take place in the walkthrough itself. On the other hand, games played by the producer, the coordinator, or other participants before and after the walkthrough can affect everyone's behavior *in* the walkthrough. So, we should discuss these games without worrying too much about when they take place in the office— they all have an impact on the interpersonal relationships in the walkthrough.

Lack of space prevents me from discussing all 35 games that Berne identifies. In addition, some of the games— for example, the sexual games— have only limited reference to our discussion of programmers and systems analysts in a walkthrough environment. But there are 8 games that I have frequently seen played in the data processing community, and these are discussed below. For more details and for a description of the other types of games, read *Games People Play* as well as Berne's other books ([Berne, 1961] and [Berne, 1963]). Also, for more recent views of the psychology of computer people, see *Techno Stress* [Brod, 1984] and *The Second Self* [Turkle, 1984].

9.3   The Alcoholic Game

The Alcoholic Game is, in the narrowest sense, a game played by a person with a serious drinking problem, with two to four other people.   In its broadest sense, though, it describes a game involving *any* kind of "alcoholic" or junky—a person who acts as if he is addicted to something.

The central player in the game is the Alcoholic, the person with the problem.   He interacts with the Persecutor, who berates the Alcoholic for his evil ways, and with the Rescuer, who tries to help the Alcoholic overcome his problems.   There are other minor players, too: the Patsy, who unwittingly provides the Alcoholic with the means to continue his bad habits; and the Connection, who provides the direct source of supply for the Alcoholic's bad habits.

The important point is that all the players derive psychological satisfaction from their roles, though the "satisfaction" may be simply a way of coping with other, deeper problems.   The Alcoholic, for example, plays the game as a form of rebellion ("see if you can stop me"); he also uses it as a form of self-castigation, and as a way of avoiding other personal and social interchanges.   The Rescuer, the Persecutor, the Patsy, and the Connection all have their stake in the game, too.

What does this have to do with data processing and the subject of walkthroughs?   Consider the data processing variations on the Alcoholic Game: the programmer who always shows up late for work; the systems analyst who procrastinates and constantly fails to meet his deadlines; the programmer who repeatedly "forgets" to attend a walkthrough; or the programmer who continually puts the same bugs into his code.   The Persecutor is often the manager, though it may be one of the programmer's peers; the Rescuer, the Patsy, and the Connection typically will be fellow programmers or systems analysts.   With the introduction of walkthroughs, it becomes easier for the Alcoholic to play his game, and it is easy for many of the walkthrough participants to fall into one of the supporting roles.

What can you do if you see the game of Alcoholic being played? It is beyond the scope of this book (and beyond my personal competence) to provide detailed suggestions for dealing with the many ramifications of this game; indeed, even a professional therapist has difficulty getting the Alcoholic to stop playing his game. However, it should be possible for you to say to your colleagues, "Hey! It looks like we're playing the game of Alcoholic; and it looks like I'm the Patsy. I don't think any of us ought to be playing this game— and in any case, I don't want to play any more."

9.4   The Game of "Now I've Got You"

The full name of this game is Now I've Got You, You Son of a Bitch. It is characterized by Player A, who discovers that Player B has made a mistake or has somehow put himself at a disadvantage; at this point, Player A becomes much more interested in the fact that he has Player B at his mercy than he is in correcting the mistake or resolving the problem.

Walkthroughs provide an excellent opportunity for this game to take place. If the producer is presenting his code or his design, one of the reviewers may suddenly leap to his feet and shout, "Aha! Gotcha! Look at that bug on page 13 of the program listing! I *told* you that you didn't understand how to use nested IFs in COBOL!"

For the person playing the Gotcha game, there are several advantages: It provides justification for his rage (rage which may have been caused by other factors), and it avoids a recognition of his own deficiencies.

The victim of the Gotcha game has little choice but to respond in a low-key, calm, objective fashion. If a bug is found in his program, he should accept it as gracefully as possible. Naturally, he may wish to point out (probably outside the walkthrough) that the game is being played, and that he doesn't wish to participate. In the best of cases, the situation can be confronted, and the game can cease; in the worst, the producer may decide not to invite the Gotcha player to subsequent walkthroughs.

9.5   The Game of "See What You Made Me Do"

This game should be familiar to everyone. Most of us will admit that we've played it at one time or another in our lives.   Typically, Player A is engrossed in some activity, working by himself; Player B comes on the scene and interrupts Player A by asking a question or making a comment. As a result of the interruption, Player A makes a mistake—he breaks the piece of furniture he was mending, he deletes all of his files from the disk by typing DELETE *.*, and so forth. Furious with the results of his mistake, he yells at Player B, "See what you made me do!"

A somewhat more subtle form of the game takes place in the walkthrough environment. Typically, the producer will object mildly to some criticism or suggestion made by one of the reviewers, but will then accept the suggestion with a show of diplomacy. Later, when the reviewer's suggestion causes the producer's program to abort and waste several hours of computer time, he will yell, "See what you made me do!" Depending on the way the game is played, the producer may be trying to avoid responsibility for making decisions about his product, or he may be trying to obtain vindication for the "mistaken" suggestions that were forced on him by the reviewers.

The other participants in the walkthrough can respond in a variety of ways. They can respond to the producer's unspoken request—"Leave me alone!"—and refuse to attend any more of his walkthroughs. Or, if they think that the producer is trying to shirk his responsibility to make a difficult technical decision, they can be sure not to get stuck with the responsibility themselves. However, note that this doesn't resolve the problem if the other participants are working on a team project whose success depends critically on the success of each individual's work. Intervention by the project manager or other higher levels of management may be required.

9.6   The Game of "Harried"

The programmer or systems analyst who plays the game of Harried is the one who takes on every assignment that is given to him, despite the fact that he is overworked. He accepts all of his supervisor's criticisms, yields to all of the demands made by his peers, and even asks for more. Since it is impossible for him to handle all of the assignments he has accepted, he falls further and further behind and begins to make more and more mistakes.

This may become evident in a walkthrough environment if the reviewers notice that Harried did not submit a complete product for review, or that he did not resolve all of the open issues that had been identified in a previous walkthrough of the same product. As a reviewer, Harried may be the one who never has time for anything other than a superficial review of other people's products. In general, Harried is the one who *always* arrives ten minutes late at the walkthrough, despite constant reminders from his colleagues.

For a variety of psychological reasons, the player may have deliberately chosen the job he is in, and he may have been attracted to his position precisely because he knew that the boss *would* continue making more and more demands on him.

Dealing with this kind of game is beyond the scope of this book, and certainly beyond the ability of most programmers or systems analysts in a typical MIS organization. If you see the game being played by one of your colleagues, there's not much that you can do to help other than to try to bring the situation to the attention of someone who is competent to deal with it. In the meantime, you have to be aware that the Harried player is probably taking on assignments that he can't handle and making commitments that he can't keep. As a matter of self-defense, you will want to ensure that you don't become an innocent victim of the catastrophes that eventually will ensue.

9.7   The Game of "If It Weren't For You"

The game of If It Weren't For You normally is considered a marital game. A wife will say to her husband, "If it weren't for you, I could have been a famous actress." Or the husband will say, "If it weren't for you and the kids, I could have been an astronaut." The game offers a tremendous advantage to the player; it allows him to avoid dealing with the reality that he could never have become an astronaut.

In the day-to-day humdrum of an MIS organization, If It Weren't For You can be played in a variety of ways. Most often, it is played by the MIS manager, who complains to his staff, either singly or collectively, "If it weren't for you, we would have this project done on time, and I would be promoted to Vice President." It is rather difficult for the programmer or systems analyst to respond to this attitude: About the only thing he can do is offer to quit, and the fact that the manager eventually will have to admit that he's never going to achieve the Vice Presidency will be a hollow victory to the programmer/analyst!

The game of If It Weren't For You also takes place among peers, and walkthroughs in the organization can be the source of the game. The producer, for example, may say to his fellow reviewers in the walkthrough, "If it weren't for you, I would have finished my program on time—but as it is, the boss is yelling at me because I missed my deadline." By blaming the reviewers, the producer manages to avoid facing the fact that his coding or functional specification was so bad that he *never* would have made the deadline.

The straightforward way of responding to such a game is to say, "We're sorry you feel that our reviews are slowing you down. Perhaps you should develop your programs yourself; we have other things to do with our time." But, as noted earlier, this doesn't resolve the problem if the other reviewers are working on a team project whose success critically depends on the individual success of the person playing If It Weren't For You. Again, outside management intervention may be needed.

9.8   The Game of "Look How Hard I Tried"

Many people refer to this game as the Martyr Game. The player is usually involved in a project of some kind—a marriage, a computer program, or a business project. He has already decided that the project is going to be a failure and wants (a) to get sympathy from his peers, (b) to be perceived as helpless to prevent the failure, so that someone else will step in and take over, and (c) to be perceived as blameless for the failure.

In the data processing field, it is easy for the programmer or systems analyst to play Look How Hard I Tried. He can stay at work all night long; he can sleep on the computer room floor while his program chews up computer time over the weekend. If he has a terminal at home, he can leave it logged on the office computer all night long, so that his colleagues get the impression that he has been working around the clock.

One indication that this game is being played in a walkthrough environment is an air of utter resignation and defeat on the part of the producer. "Yes, you're right about those bugs," he'll say. "I don't know how they got in there after I worked through the last two weekends, and Christmas, too, trying to get it all right." Another indication is that the number of open, outstanding issues for the same product remains relatively constant, even after the third or fourth walkthrough. Each time, the producer is likely to shrug his shoulders and say, "Gosh, I tried as hard as I could. I stayed until midnight last night trying to get those bugs out..."

In most cases, the game is played between the programmer or systems analyst and his manager. The other programmer/analysts are pawns in the game who are manipulated by the player to help emphasize to the manager just how hard he was trying. Obviously, the only thing they can do is to try to stay out of the game, and let the player and the manager resolve things for themselves.

9.9   The Game of "Schlemiel"

The player in this game— the Schlemiel— manages to make a number of mistakes, all at the expense of his victim(s). In the data processing community, the Schlemiel accidentally spills coffee all over his officemate's program listing, or even worse, all over the keyboard of his officemate's personal computer or floppy disks; he inadvertently deletes all of his colleague's files from the library, and so forth. The victim initially feels anger or resentment at these mistakes, but feels that if he shows that anger, the player will "win." So, he swallows his rage and smiles at the Schlemiel. When the Schlemiel apologizes, the victim, if he follows the rules of the game, offers forgiveness. The game continues, often over a period of days or weeks. The Schlemiel continues to make mistakes, until finally the victim no longer is able to contain his rage.

You can imagine a number of examples of this in a walkthrough environment: The Schlemiel forgets all about his own walkthrough and fails to attend. Or he fixes one trivial error pointed out in the previous walkthrough but in so doing, manages to introduce three new fatal errors.

How does one deal with this game? In a sense, the victim can't win: If he shows anger at the first incident, the Schlemiel will feel justified in returning the resentment. If he swallows his anger, the game will continue indefinitely. The best approach is the direct one and one that most victims find difficult to put into practice. The victim might say, for example, "Wow, you really *are* a Schlemiel! I guess I'd better change my password so you don't destroy any more of my files. And it's really a pain in the neck to have you spilling coffee on my personal computer— maybe I should talk to the boss about getting our office assignments changed."

9.10   The Game of "Yes-But"

The last game we'll discuss is one that comes up often in a walkthrough environment. The player begins the game by stating a problem; for example, "I don't see any way of coding this part of the program without using nested IF statements, but our standards say that we're not allowed to use nested IFs." One of the other participants responds by offering a solution, such as, "Why don't you solve the problem by using a decision table?"

Unfortunately, every solution that is offered is rejected with a "Yes, but..." comment by the player. For example, "Yes, but the decision table preprocessor generates terribly inefficient code." If another participant offers a different suggestion, it too will be rejected. Thus, if he says, "Well, why don't you code the program with an ELSE-IF, or CASE-like, construct instead of nested IFs?" the response will be, "Yes, but that kind of code looks ugly." And so it goes.

Eventually, it becomes obvious that *no* solution will be accepted by the player. The way to get out of this game is to say to the player, "Yes, it *does* seem to be a difficult problem! How do you propose to solve it?" However, in keeping with the suggestions in Chapter 6, any lengthy discussions of this kind should take place *outside* the walkthrough environment, in order to avoid wasting the time of all the other participants.

9.11 Summary

We have looked briefly at only a few of the more common games that take place among people. In certain social situations, such as parties, the games become pastimes and may be a pleasant way of whiling away the time. In a situation like a walkthrough, though, the games interfere with the immediate business, and they can lead to a complete breakdown of communication among the various participants.

The primary objective of this chapter is to make data processing people aware that such games *do* exist. Prior to the introduction of walkthroughs, they may not have been forced into sufficiently close contact with their colleagues to have to face the reality of games. As we have seen in many of the examples, the best solution is to avoid getting sucked into the game—or to confront the player directly.

QUESTIONS FOR REVIEW AND DISCUSSION

1. Can you think of examples of each type of game that has been discussed in this chapter? Discuss them with your colleagues and see whether you all agree on the proper way of dealing with the games.

2. Can you think of additional types of games that take place in your walkthroughs? If so, consult *Games People Play* to see whether they are standard games.

3. Do you agree that the best way of dealing with most of the games is to refuse to participate? Are there any other solutions that you can see?

10 JUSTIFYING WALKTHROUGHS TO MANAGEMENT

> The presence of a body of well-instructed men, who have not to labor for their daily bread, is important to a degree which cannot be overestimated; as all high intellectual work is carried on by them, and on such work material progress of all kinds mainly depends, not to mention other and higher advantages.
>
> Charles Darwin
> *The Descent of Man*, Chapter 4 (1871)

In some organizations, data processing managers are unconvinced that walkthroughs are a cost-effective way of doing business. "What a waste of time," they say, "letting a bunch of programmers sit around a table and argue about their own programs."

Or a manager will say, "Why not let the computer find all those bugs? After all, we have on-line program development facilities, so the programmers can find their own bugs at their own workstations! Isn't that why we got all those PCs and CASE tools in the first place?"

There is often some validity to these comments, and even if there weren't, many managers would *think* that there is. Thus, these objections need to be discussed. Are walkthroughs *really* cost-effective? In all cases? Can the cost-effectiveness be demonstrated? Is it tangible? Or does the manager have to rely on some gut feeling that the intangible benefits of walkthroughs make them worthwhile?

My experience, and the experience of a majority of my data processing colleagues and MIS clients, is that walkthroughs are extraordinarily worthwhile and their benefits can be measured in tangible terms in most cases. The purpose of this chapter is to discuss the benefits of walkthroughs from a management viewpoint, so that managers *can* be convinced that the investment in walkthroughs is worth the effort.

10.1   The Economics of Bugs

One of the major purposes of a walkthrough—if not *the* major purpose—is to find errors or bugs in a product. From a management viewpoint, the question should be: Can bugs be found more economically in a walkthrough environment than in the classical environment? Since the classical environment usually requires the producer to find his own bugs prior to a formal review, the real questions are: Is it more economical to let the producer find his own bugs, or is it better to let a group of three to five people find the bugs? Or, in a more advanced MIS development environment, is it more cost-effective to let the developer use automated tools (testing packages, simulators, etc.) to find the errors?

The answer is, *it depends*. If the producer is a genius (someone whom my colleague Nan Matzke refers to as "the old reliable") who never makes mistakes, a walkthrough is almost certain to be a waste of time and money.[1] However, if the producer is an idiot who can barely spell his name without making a mistake, walkthroughs are well worth their time and expense. But most producers are neither idiots nor geniuses, nor are the reviewers who participate in the walkthrough. So it is not intuitively obvious how the manager can eliminate bugs in the most cost-effective manner. As Barry Boehm points out in *Software Engineering Economics* [Boehm, 1981], "a great deal of important work remains to be done in this area in order to obtain a thorough understanding of the relations between software cost and software reliability."

---

[1] Even in this case, the walkthrough has an important side benefit: It can be used to help train the next generation of geniuses, and it can provide insurance in case the current genius decides to move on to a new job.

One factor in this cost-benefit calculation is the *turnaround time* for fixing errors. Originally, this was primarily a concern for code walkthroughs, for programmers often worked in a batch computer environment, with overnight turnaround. Today, most programmers develop their code in an on-line environment, often with their own PCs or workstations, and they typically expect near-instantaneous turnaround for compilations and test shots. But more importantly, system developers now realize that coding is only a small part of the job of developing information systems, and we now worry about turnaround time for other phases of the project.

Turnaround time has an entirely different significance when we consider the activities of systems analysis. The graphical models of dataflow diagrams, entity-relationship diagrams, etc., are typically drawn by hand in most MIS organizations, and the time required to redraw them can range from a few hours to a few days. Of course, if automated CASE tools are used, revisions can be made in a matter of minutes, but this technology will not be common in most MIS organizations until the mid-1990s. More important is the turnaround time required to schedule the next meeting between the systems analysts and the users to determine whether there are any additional errors or misunderstandings in the specification; this can range from days to weeks, depending on how widely dispersed the user community is.

What does the programmer or systems analyst do in the meantime? Whether the manager wants to admit it or not, the answer quite frequently is: *nothing*. Of course, the programmer/analyst will *look* busy; however, it's likely that he will be doing busywork that serves very little purpose.

Sometimes the programmer/analyst does have other worthwhile things to do: other modules to code, other parts of the system specification to develop, other walkthroughs to attend, etc. But still, time is lost whenever he has to switch gears mentally from one task to another. And the longer he has to wait before getting back to work on the original task, the more difficult it will be to remember what he was doing.

Thus, one of the objectives of a walkthrough is to ensure that the product is correct the *first* time. Thus, if the walkthrough is successful, a program should only have to be compiled *once*; it should only have to be tested *once*.[2] And it should only be necessary to have one formal presentation of the system specification to the user community, because the specification walkthrough (which should include one or more user representatives) would have found all of the errors in the specification.

To illustrate this, let's examine a specific example. I once had the opportunity to participate in a walkthrough with three programmers in a small programming shop. The four of us spent approximately 45 minutes reviewing a program composed of roughly 200 PL/I statements. The investment of three work-hours paid off handsomely: We found a total of 25 bugs. It's particularly interesting to see what *kinds* of bugs they were:

- approximately 5 bugs were syntax errors

- approximately 5-10 bugs were trivial logic errors

- approximately 5-10 bugs were moderately complex logic errors

- approximately 5 bugs were "impossible" logic errors

Why five syntax errors? The reason is simple: The walkthrough was conducted immediately after the program had been written, but *before* it was entered into the computer, and *before* it was compiled. As I first observed in Chapter 2, walkthroughs are usually not conducted at this early stage. The producer's writing was hard to read, and the hand-written PL/I statements had been reproduced on an ancient copier that produced nearly illegible copies on a waxy

---

2 That is, the program should only have to be tested once, with oodles of test cases, to see whether it works by itself. One would expect to see the program tested several more times in a system testing environment, as part of integration testing.

paper similar to the toilet paper one often finds in Parisian bathrooms.

It was not pleasant reading the program under those conditions. Nevertheless, it was probably cost-effective to find those five syntax errors, even though the PL/I compiler would have found them in a matter of milliseconds. Because the organization had overnight turnaround for compilations (the same computer was used for production runs and for program development, so program development got relegated to the third shift), a day would have been lost while the programmer fixed the syntax errors and resubmitted the program for compilation. Further, there is no guarantee that she would have fixed those errors without introducing a few new ones.

As for the logic errors, it is absolutely certain that the walkthrough was more cost-effective than allowing the producer to find them by herself. Several of the bugs were trivial and would have been exposed with any kind of test data. Nonetheless, they would have required at least one recompilation of the program, so at least one more day would have been lost. Indeed, it's much more likely that several recompilations would have been necessary, especially since the producer was the kind of programmer who typically would find only one bug with every test run.

Several of the bugs would have been even more expensive had the walkthrough not found them first. It would have been evident to the producer that something was wrong because, for example, the program would have blown up or would have produced gibberish output. But it would not have been intuitively obvious *what* was wrong. Left to her own devices, the producer would have used program dumps, traces, and other debugging techniques to isolate the bugs, all of which is time-consuming and expensive, not only in terms of the producer's time, but also in terms of computer time, printer paper, and computer operator's time.

Finally, there were the "impossible" bugs— some five situations which *never* would have been found by the producer. Since the program was written in an organization that was too small to have a separate Quality Assurance Department or Testing Department, it follows that these bugs would have remained in the code until it was put into production. A friend of mine calls these bugs "time bombs," and the description is an apt one!

Most walkthroughs don't produce results quite as dramatic as the example described above; it is more common to find three or four errors in a walkthrough. Even so, this detection means that the program can be compiled, tested, and made ready for further system testing in one fell swoop, and the programmer doesn't waste several days waiting to get the bugs out of his code.

The payoff tends to be even greater when bugs are found in design walkthroughs and walkthroughs of user requirements. A classic study cited by Barry Boehm [Boehm, 1981] showed that it is roughly ten times less expensive to fix design errors that are detected in the design phase of the project than it is to fix them after the code has been written. In addition, this means a product that has had walkthroughs in the earlier stages of development can go into production without the usual last-minute problems and crises that one often expects to see in the development of a large, complex system.

Similarly, a study by Michael Fagan [Fagan, 1976] at IBM found that inspections (a more formalized, ritualized review process than the walkthrough approach discussed in this book) improved productivity by 23 percent. And studies by Jones [Jones, 1977], Boehm [Boehm, 1980], Crossman [Crossman, 1979], and Myers [Myers, 1978] indicate that *code* walkthroughs can remove between 38 percent and 89 percent of the errors, and that this can be done at a speed of 10-30 lines of source code per person-hour. Finally, a retrospective analysis by Thayer [Thayer, Lipow, and Nelson, 1978] of several thousand problem reports in large TRW software projects found that inspections of the logic (as expressed in flowcharts, HIPO diagrams, pseudocode, etc.) of individual modules could find 58 percent of the errors; inspections of the actual code (expressed in COBOL, Pascal, Ada, etc.) could find 63 percent of the errors.

Having stated all of this, let's mention some situations where the walkthrough might *not* be cost-effective:

1. If the programmers have access to on-line program development facilities with fast response time, then some of the arguments above might not be as valid. Even so, the computer will only "find" the bugs that the producer is clever enough to expose with properly chosen test cases; and debugging, even in an on-line environment, can be a time-consuming process. In such an environment, I suggest that the walkthrough take place *after* a program has been compiled (with no remaining syntax errors) but *before* the producer has spent any time doing his own testing. Note that the same comment applies to design walkthroughs and specification walkthroughs: If the producer has access to automated CASE tools with which to produce a specification or design model quickly, then a walkthrough should take place after a legible document has been produced, and after the automated error-checking facilities have found all of the straightforward syntax errors.

2. If the producer is known to be a near-genius who rarely makes mistakes, the walkthrough may not be necessary. *Very* few managers would be willing to bet their next paycheck that their systems analysts and programmers could produce a complex system without errors, but one occasionally does run into "simple" projects that might be produced error-free by an average systems analyst or programmer. A compromise might be to insist that even the genius' work be reviewed by at least one other peer. Once again, with appropriate office automation facilities, this process could be accomplished quite easily *and controlled directly by the project manager.*

3. If the *cost* of a bug is small enough, a walkthrough may not be cost-effective. It has been traditional in some organizations to let the user find the last few

bugs in a computer system after the system is put into production, *presumably because the organization could afford to let them find the bugs.* In most of today's large, complex systems, we no longer can afford to do business this way. A bug in a large on-line system, for example, often costs the organization thousands or even tens of thousands of dollars.[3]   But, as a manager, you may wish to evaluate the cost of discovering a bug in a production system versus the cost of discovering it in a walkthrough. In his classic book on testing, *The Art of Software Testing* [Myers, 1979], Glen Myers cites studies showing that code walkthroughs are effective in finding 30-70 percent of the logic and design errors in typical programs.   Code inspections at IBM, he points out, have been effective in finding as much as 80 percent of the coding errors. As he states:

> human processes tend to be more effective than computer-based testing processes in finding *certain* types of errors, *while the opposite is true for other types of errors.*

The implication is that inspection/walkthroughs and computer-based testing are complementary; error detection efficiency will suffer if one or the other is not present.

4.   Walkthroughs may be less cost-effective at catching certain kinds of errors than other defect-removal techniques such as classical unit testing. It would be very tedious, for example, to look for errors in numerical approximations in a walkthrough—e.g., reviewing a module whose job is to compute sines and cosines to eight digits of accuracy. And it is often difficult to spot subtle errors involving the "dynamics" of a program's execution—e.g., timing errors, "race" conditions, queue overflows, etc. On the other hand, as my colleague Mark Wallace points out, it's often even more difficult to find such

---

3 For an interesting discussion of the cost of software bugs, see [Neumann, 1985].

things as timing errors via normal testing, because it is so difficult to control the timing of the input test cases; in the absence of highly sophisticated simulators and emulators, walkthroughs may be the *only* effective way of finding such bugs.

10.2   The Psychology of Finding Bugs

In many discussions about walkthroughs, there is an assumption that the producer could find the bugs in his program or design if he were given enough time. But in the example I discussed earlier, I mentioned the concept of "impossible" bugs—bugs that the producer would *never* find, regardless of his efforts.

In looking for errors in his product, the producer is likely to make the same logical errors that he made when he built the product in the first place. If, for example, a programmer forgot to check for an exceptional situation in his code, then he'll probably forget to invent a test case which will cause the exceptional situation to occur.

Many managers argue that the problem is even more fundamental. The producer has no psychological motivation to invent test cases that will demonstrate that his product is flawed or wrong. Instead, he is motivated to demonstrate that his product works! One can see this attitude among programmers quite often. They treat their programs like fragile pieces of china, and will provide only those test cases that are sure not to upset the delicate mechanisms within the code. Given this state of affairs, it is not so surprising that a number of bugs will remain undiscovered forever if the testing is done exclusively by the producer. Forever, that is, until the product is delivered to the end user, at which time the cost of discovery and correction of the bug is almost certain to be much higher than the cost of a walkthrough.

10.3   Intangible Benefits of Walkthroughs

In Chapter 1, I pointed out three intangible benefits of walkthroughs:

- improved quality of the product

- training and exchange of technical information among the programmers and systems analysts who participate in the walkthrough

- insurance: a greater probability that the product can be salvaged if the producer leaves the project before he is finished

How much are these benefits worth? In the final analysis, only the data processing manager can tell. All I can do in this book is state emphatically that the majority of MIS managers with whom I have talked during the past ten years have confirmed that the walkthroughs *did* improve the quality of their product, and that they *did* increase the level of technical expertise on the part of the staff, and that they *did* eliminate the need for scrapping partially completed products in the event of staff turnover.

There is one other reason for the manager to introduce walkthroughs in his organization. They make day-to-day work more fun for most programmers and systems analysts. There are some cases, of course, where programmers and systems analysts complain that walkthroughs are not fun (sometimes because of the psychological games discussed in Chapter 9), but most agree that it's much more fun to work in an environment where problems, discoveries, excitements, and all of life's day-to-day trivia can be shared with one's peers. In simple terms, many managers have reported that the morale and general "spirit" of their staff have improved significantly as a result of walkthroughs.

10.4   Situations Where Walkthroughs May Not Be Justified

As I have already pointed out, there may be some situations where an MIS manager cannot cost-justify a walkthrough. If the producer develops high-quality, error-free products, or if the cost incurred by a bug is small, there may not be much incentive to subject the product to a walkthrough. Since the walkthrough probably costs between

$150 and $1,000 (assuming 1-2 hours of time invested by 3-5 people whose salary, including 100 percent overhead, probably averages $50-100 per hour in 1988 dollars), the manager has to ask himself whether he can expect to receive $150-$1,000 in benefits.

There are one or two common cases when a manager may feel that it is difficult to justify a walkthrough. One such case is the one-person project: a project worked on by a single producer for a short period. Other programmers or systems analysts in the organization may be working on entirely different one-person projects, so no one can be expected to have any familiarity with the producer's product. In addition to the normal investment of time required by the walkthrough, the manager has to ask himself whether it is worth spending the time allowing the producer to familiarize his fellow programmers and systems analysts with the nature of the application on which he is working. This step might require another hour or two and involve specialized knowledge that the other programmers/analysts do not have. I have seen this situation primarily in small data processing organizations—for example, a small scientific data processing organization where the programmers write relatively small FORTRAN programs for engineers, chemists, physicists, and other scientists; or a business data processing group where the programmers use report generators or fourth generation programming languages to produce simple reports for managers in other parts of the company. If each programmer and systems analyst is working on distinctly different applications, then it may not make sense to require them to walk through each other's products. However, the situation is usually not so extreme. The applications are frequently similar (after all, a report is a report regardless of the details, especially if it is produced by a fourth generation language), and it often takes only a few minutes to familiarize someone with the general nature of one's product.

However, very few one-person projects remain one-person projects forever. Sooner or later, the producer moves on to bigger and better things, and his computer system is turned over to someone else. Without a walkthrough, the product could be relatively unintelligible to the maintenance programmer. This problem is particularly severe in the case of user-developed software, since the "maintenance programmer" is usually just another end user, without any special programming skills.

A manager may also question whether or not it makes sense to have walkthroughs in a maintenance environment. Once again, it's a question of trade-offs. If the maintenance activities involve making one-line changes in small programs, and if the cost of a mistake is small, then there's no urgent need to insist on walkthroughs. On the other hand, if the product is large, such as a system containing several hundred programs and several hundred thousand lines of code, a walkthrough is likely to have the benefits that we have discussed throughout this book, especially if the original product was developed in an "unstructured" fashion several years ago. What appears to the maintenance programmer to be an innocent change to a "rat's-nest" system may have subtle ramifications throughout other parts of the system. Walkthroughs thus have the advantage of minimizing the nasty surprises of maintenance programming.

In some maintenance organizations, walkthroughs have come about for a different reason: The staff is too small to assign each programmer his own program to maintain. One of my clients, for example, has a staff of three programmers who maintain some five hundred different programs. Most of these work well, which is lucky since the three programmers couldn't possibly manage five hundred "buggy" programs! But they don't know *which* program is likely to cause them trouble at any point; or, to put it another way, they don't know which of their programs will require modifications and enhancements to suit the whim of a user. Also, of course, it's impossible to remember the details of five hundred different programs over a long period of time, especially if you look at them only once or twice a year. Consequently, the three programmers work as a team. If a program has to be changed, all three programmers will look at the listing, discuss the change, and walk through the modified code. The whole process may occupy them for only an hour or two, after which they go on to the next program to be modified.

Finally, the manager may have to abandon the idea of walkthroughs if his project is seriously behind schedule. There is an old saying that applies to MIS development projects: "People never seem to have time to do a job right the first time, but they always have time to do it twice." Unfortunately, this saying is often true of MIS projects that were estimated originally to take one year but then stretch into two years. Pressure from impatient users and frustrated higher-level managers means that systems are designed hastily, coded sloppily, and tested inadequately. We now can add an item to that list of sins: Such projects find that their walkthroughs are done hurriedly, without adequate preparation— or not at all.

10.5 What Can Go Wrong with a Walkthrough?

So far, I have addressed problems that might prevent a walkthrough from being held, but I have implied that once a walkthrough begins, it will be successful. Obviously, there are no guarantees that it will be. As I observed in Chapter 6, walkthroughs are *not* a magic ritual that causes bugs to rise out of the program listing and smack the reviewers on the nose. It is possible that the programmers and systems analysts will fall sound asleep in the walkthrough, and nothing will happen. Thus, the manager must ask: What can go wrong in the walkthrough? How can the problems be prevented?

The major problems, in my experience, have been as follows:

- The participants goof off and begin discussions that have nothing to do with the product being reviewed.

- The participants become involved in long arguments about minor, if not trivial, points; arguments about the style of the product are especially common.

- The participants become involved in personality clashes which obscure the real subject of the walkthrough.

These problems, if not properly controlled, can render the walkthrough useless. Even worse, the participants can become embroiled in such bitter arguments that they may stop speaking to one another.

To avoid these problems, the manager must ask himself whether the following assumptions are valid about the people who participate in a walkthrough:

1. They are inexperienced in the procedures and mechanics of the walkthrough.

2. They are human and are thus prone to lapse into discussions of irrelevant subjects, even though they know they shouldn't.

3. They are inexperienced at giving or taking criticism, so they may have trouble discussing bugs or flaws in the product without making someone (usually the producer) lose his temper.

4. On the other hand, they basically respect the intelligence, competence, and integrity of the other participants in the walkthrough.

5. They feel a sense of responsibility for making the product succeed within the given constraints— e.g., schedule, budget, and so forth.

If assumptions 1, 2, and 3 are incorrect, then the participants have little excuse for wasting time or conducting an unsuccessful walkthrough. However, if assumptions 4 or 5 are incorrect, there is virtually no point in holding a walkthrough because it is almost guaranteed to fail.

This last point is a serious one, especially if the participants in the walkthrough are chosen arbitrarily (that is, according to some corporate standards) or if the participants are chosen by an amateur (such as the producer himself, who may not appreciate the human dynamics of the situation). Thus, the most important thing that the manager can do is to ensure that the participants do respect one another, and that they *do* feel a sense of responsibility for the success and quality of the product.

Once he has done this, there are several specific guidelines for overcoming the problems listed above:

1. *Have a written set of procedures for conducting the walkthrough.* This book provides a default set, tested by experience. Use it as a starting point, and customize it for your own organization.

2. *Be prepared to let the participants waste some time.* Remember that they aren't experienced in the mechanics of walkthroughs, so the first one or two walkthroughs may be an utter waste of time. Be prepared to walk past the conference room where the walkthrough is being held and hear raucous laughter, shrieks, and giggles, and all the other evidence of a drunken orgy. Don't be upset if the group orders pizza or hamburgers to accompany the walkthrough; they're only trying to be human.

3. *Rely on the group's sense of responsibility.* After one or two unproductive walkthroughs, the participants almost always will begin worrying about their lack of progress if they have a sense of responsibility toward the product. If they get into the same argument that they got into on the previous walkthrough, one of the participants is likely to say, "Hey, we got into this argument last time—we're wasting time going over this again." A compromise reached this way is much more effective than if the manager forces a compromise.

4. *Enforce the thirty- to sixty-minute time limit.* One of the primary reasons for chaos in a walkthrough is that the coordinator allows the discussion to get out of hand. Typically, this means that the walkthrough

drags on for an hour or two, with the participants arguing the same points over and over. While the manager should not take over the coordinator's role in the walkthrough (see Chapter 11 for more details on this), he sometimes can help impose discipline by strictly enforcing the thirty- to sixty-minute time limit. If the participants know that they have to be finished and out of the conference room at the end of an hour, they will tend to stop their bickering soon enough to get some work done.

5. *Insist that the participants sign the walkthrough report.* As I discussed in previous chapters, the standard walkthrough procedure requires the participants to sign the walkthrough report after voting on the status of the product. Some participants think this step is a formality and will conveniently forget to do so; the coordinator may also be lax in asking the participants to sign the report. From the manager's point of view, though, getting the signatures is important, for it emphasizes to the participants that they *are* responsible for the outcome of the product. One manager friend of mine solved his problems with walkthroughs by insisting that the walkthrough reports be signed, and then taking all of the reports home with him each night. If any of the computer systems aborted in the middle of the night during a production run, the computer operator was instructed to call the manager; the manager then called all of the participants who signed the walkthrough report of the aborted program, and dragged them from their beds into the computer room to fix the bug. The message was clear: The author of the program may have been to blame for creating the bug, but the reviewers were equally guilty for not finding the bug during the walkthrough.

6. *Make sure that the participants have some standards for programming, design, and systems analysis.* As I have mentioned, many of the

arguments that occur in a walkthrough concern the style of the product. These arguments can be reduced, though probably not eliminated, by ensuring that the participants have a set of standards with which they can all agree. In some cases, the standards will be a subset of the corporate standards; in other cases, the group may want to adopt a superset— stricter than the normal standards. In any case, the manager may find that the first few walkthroughs in a new development project will concern the standards by which the group agrees to design and code. Once these are agreed upon, subsequent walkthroughs should run much more smoothly.

7. *Make sure that participants realize that participation in the walkthrough is part of their job, and that their performance review will reflect management's assessment of their participation.* If participants feel that management doesn't care about walkthroughs, or doesn't care whether team members participate in them, then there is always the danger that some programmers and analysts will regard them as a waste of time.

QUESTIONS FOR REVIEW AND DISCUSSION

1. Do the managers in your organization feel that walkthroughs generally are cost-effective and worthwhile? If not, what kind of arguments could be used to convince them?

2. As an experiment, try to estimate the average cost of a walkthrough by computing the average salary of the participants, plus whatever overhead factor is normally used by your organization. Then try to estimate the number of bugs that will be found in the walkthrough and the savings that are likely to result— i.e., estimate how much money would have been spent finding the bugs in the classical fashion. Can you demonstrate the effectiveness of walkthroughs in this manner?

3. What is the average turnaround time for compiling and testing programs in your organization? What do the programmers *really* do while they're waiting?

4. How many bugs would you expect to find in a typical walkthrough?

5. If you have access to a fast-response on-line program development system, do you think that walkthroughs are worthwhile?

6. Do you agree that producers find it psychologically difficult to find their own bugs? Can you cite any evidence of this difficulty in your own organization?

7. Are walkthroughs worthwhile in an organization heavily oriented toward one-person projects? Why or why not?

8. Are walkthroughs worthwhile in a maintenance environment? Why or why not? Is the situation fundamentally different from a development environment?

9. What should a manager do if he finds that the participants in a walkthrough are goofing off and not devoting their attention to the subject of the walkthrough?

10. What should the manager do if he finds that certain participants have serious personality clashes with one another?

11. Do you agree that the manager should allow the participants to "waste" the first one or two walkthroughs? What are your reasons?

12. Why is it important that the participants be asked to sign the walkthrough report?

11 AVOIDING THE URGE TO ATTEND WALKTHROUGHS

You do not need to leave your room. Remain sitting at your table and listen. Do not even listen, simply wait. Do not even wait, be quite still and solitary. The world will freely offer itself to you to be unmasked, it has no choice, it will roll in ecstasy at your feet.

Franz Kafka
The Great Wall of China

Throughout this book, I have advised that managers *not* participate in walkthroughs conducted by their subordinates. However, I have only hinted at the reasons for this advice, and while some people may feel the reasons are obvious, others may be thoroughly puzzled. Hence this chapter, whose purpose is to explore the issue more deeply.

I will begin by discussing the reasons that a typical programmer or systems analyst feels intimidated by the presence of a manager in a walkthrough. Then I will examine some of the reasons that a manager *wants* to participate in a walkthrough. We will see that in most cases, the manager can accomplish the things that he or she needs to accomplish without interfering with the normal walkthrough process.

11.1   Why Should Managers Avoid Walkthroughs?

The primary reason for suggesting that managers avoid attending walkthroughs is that their presence interferes with a frank, open exchange of views among peers. Regardless of what the manager actually does in the walkthrough, his mere presence usually is perceived as a signal that the *producer*, rather than the product, is going to be evaluated. As one programmer told me, "Every time the boss attends the walkthrough, I start worrying that he's keeping score in his little black book—and at the end of the year, he's going to tell me that I'm not getting a raise because I had a total of 378 bugs during the year." Even if the manager is not keeping score, it's difficult to avoid the normal human reactions when bugs are observed in the product. The producer feels a combination of fear, humiliation, and inadequacy; the manager feels a combination of condescension and impatience.

If the producer feels this way, it is difficult for him to avoid playing a variety of games in the walkthrough; in many cases, the participants respond with their own games. The producer, for example, will tend to defend his product against the criticisms and suggestions made by the participants: In the extreme case, what they perceive as a bug in his product, he will regard as a "feature" of the product. When the reviewers suggest improvements to the style or organization of the product, the producer is likely to proclaim immediately that the changes are not possible, or that they're stupid. Of course, the producer *should* be accepting such suggestions without comment and then deciding *after* the walkthrough how to deal with the suggestions. But nobody wants to make a fool of himself in front of his boss, so the producer will tend to get into arguments in the walkthrough, which means that the coordinator will have a difficult time maintaining his role.

Meanwhile, the participants may have their own games to play. If they don't think highly of the product, what better forum for voicing their displeasure than in front of the boss? More importantly, if they don't get along well with the producer, the reviewers will tend to be more than normally critical of the product. They, too, will begin to argue over issues of style when they should be simply stating the criticisms and moving on to the next point. Or, as Nan Matzke has observed, the other participants may decide to keep their mouths shut, in which case the walkthrough degenerates into a one-on-one session between the producer and the boss.

All of these reactions can occur even though the manager sits quietly in the corner of the room. In many cases, the manager is unwilling or unable to play such a passive role: He, too, begins looking for bugs in the product and begins arguing about its style. Since he is the boss, the other participants (not to mention the producer and the coordinator) are somewhat obliged to let the manager have his way. In effect, the manager ends up running the walkthrough, and because he is likely to have a stronger personality than the participants, he ends up dominating the walkthrough.

In some organizations, simple peer-group reviews have evolved (or degenerated, depending on your point of view) into reviews organized and run by the manager. The reviews are often held in the boss's office, just so that no one will forget who is in charge. If the participants, including the manager, are honest, sincere, egoless people with no ax to grind, such a review can run smoothly. In most organizations, though, one of two things will happen: (a) the review will deteriorate into a shouting match, or (b) the participants will perceive that it's best for them to keep their mouths shut and let the manager do whatever he wants to do.

Some managers don't believe all of the above. "I get along perfectly well with my programmers and systems analysts," they say. "My presence doesn't affect them at all in the walkthrough— after all, we're adults, not children!" I have noticed that such a manager rarely asks his programmers and analysts what they think about his presence; and if he does ask, the question is often posed in such a way as to make it obvious to them that they should not object to their manager's presence. If such a manager participates in a walkthrough, he's likely to find, if he looks, that (a) he has become a player in the game between the producer and the participants, or (b) the games are indeed being played, but they have been repressed for the sake of social etiquette, which is, in itself, a complex game. Or he may find that the programmers and systems analysts are not arguing with each other because they have abdicated responsibility for the product to the manager.

There is one case when it *does* make sense for a manager to participate in a walkthrough. If the manager also functions as a programmer or systems analyst himself, then his products will presumably be reviewed along with everyone else's. This case usually happens when the manager is a team leader, lead systems analyst, or senior programmer—a first-level manager who has responsibility for the day-to-day activities of a few subordinates, but who does not have authority to hire, fire, or grant salary increases. Since such a person usually is a player-coach who is producing as much technical work as his subordinates, his presence in the walkthrough is almost unavoidable. Nevertheless, such a manager should be aware that he still will be perceived by the other programmers or systems analysts as a boss. Even if he can't determine the amount of their raise, they worry that he will pass negative recommendations to the next level of manager. I will suggest a way out of this dilemma in the next section.

11.2   Why Do Managers Want to Attend Walkthroughs?

Having pointed out the problems caused by a manager's presence in a walkthrough, we should ask why a manager would want to attend a walkthrough. When we understand the reasons, we can think of ways of satisfying the manager's legitimate needs without destroying the walkthrough concept.

Though he may not admit it, curiosity is one reason why a manager wants to attend a walkthrough. He may simply wonder what goes on in the walkthrough. What do the programmers do to one another? Is it really worthwhile? What he really would like is an opportunity to watch the participants in a walkthrough through a one-way mirror; failing that, the manager will sometimes insist on attending one or two walkthroughs, just to satisfy his curiosity.

If this is the case, the participants can satisfy the manager's needs by arranging a "demonstration" walkthrough. Depending on the situation, the participants may want to select a product that is known to be in reasonably good shape, without any embarrassing bugs, or a product that was developed by someone who is no longer with the organization, or a program that already is in production. It also may be interesting to walk through a product developed by the manager back in the days when he was a technician—but watch out for the obvious political problems!

In most cases, the manager will want to sit through only one or two such walkthroughs, so it needn't be too much of a strain for the participants. Eventually, the manager will discover that he really doesn't understand the subject matter being reviewed; or he'll become bored listening to discussions about detailed technical issues; or he'll decide that he has more important things to do than attending a walkthrough.

Why else would a manager attend a walkthrough? Some managers insist on attending the walkthrough because they are convinced that the participants will goof off or become involved in endless technical arguments. In blunt terms, these managers don't trust their subordinates to conduct an effective walkthrough without supervision.

In Chapter 10, I discussed a number of ways to avoid the dangers of arguments among the participants, goofing off, and walkthroughs that stray from the subject. (Note that many of these problems indicate that the coordinator has failed in his role.) These problems can be overcome by the manager, but, as I pointed out, it may take time, and the manager should be prepared to "waste" the first few walkthroughs. If the manager is convinced, though, that the participants will never be able to conduct themselves properly in a walkthrough, then he might as well abandon the walkthrough approach completely and substitute a formal review in its place. In such a case, it would be better to have a one-on-one review between the manager and the producer—not a pleasant prospect, but probably better than having a roomful of squabbling technicians arguing about trivia.

There is yet another situation when the manager may feel strongly that his presence in a walkthrough is required: if he has considerably more technical knowledge and experience than the producer or any of the other participants. For example, the manager may have several years of programming and systems analysis experience, while his staff consists primarily of junior programmers. Naturally, the manager is nervous that the producer and the participants may overlook serious bugs or may not be aware of more sophisticated solutions to the problem.

The solution is to let the manager attend the *second* walkthrough. Let the producer and the participants have a walkthrough by themselves first. They will find many of the "stupid" errors that they wouldn't want the manager to see. Once they're satisfied with the product, then the manager can conduct a more formal review, just as a Quality Assurance Department may insist on its own review after the walkthrough has taken place.

There remains one reason for the manager to participate in the walkthrough: He may be a technician himself. As I mentioned above, many organizations have player-coaches who write programs and also supervise the day-to-day activities of other programmers. It is inevitable in such circumstances that the manager will participate in the walkthroughs of his subordinates. To minimize the political games that I warned about at the beginning of this chapter, he should be properly humble about walkthroughs of *his* products. If the other programmers and systems analysts see that the manager makes mistakes and is able to accept criticism, then they are prone to be more willing to do the same themselves.

In some situations, though, it would be better to keep the manager out of walkthroughs of his own product, because his presence will be so intimidating. As a compromise, let the manager sit in a quiet corner (*not* in the front of the room or some other dominant position) and do *not* let him present his own product. This provides at least a modicum of encouragement for the other participants to voice their honest opinion of the product.

Unfortunately, this stance won't eliminate all of the problems. The manager should expect that his subordinates will continue to be nervous about his presence in the walkthroughs for several months— nearly six months in one organization I visited. Only after he has survived one or two salary reviews is a typical programmer or systems analyst likely to *really* believe that management is not keeping a count of the number of bugs discovered in the walkthroughs of his programs.

QUESTIONS FOR REVIEW AND DISCUSSION

1. Is it standard practice in your organization for managers to attend walkthroughs?

2. Do you agree that a manager's presence is likely to make the producer feel nervous and defensive? Will the programmers and analysts openly admit that they feel this way?

3. What kind of games have you seen producers play when a manager attends one of their walkthroughs?

4. What kind of games have you seen managers play when they attend walkthroughs?

5. What can the producer and the participants do if the manager *insists* on attending the walkthrough? What do you think of the idea of having a semi-secret "private" walkthrough that the manager doesn't know about (the local pub is often a good place for this)?

6. Why do *you* think that managers want to attend walkthroughs? Are their reasons legitimate?

7. What can be done if the manager is simply curious to see how a walkthrough operates? Does the idea of a demonstration walkthrough make sense in your organization?

8. If the manager is also a producer (i.e., he writes code or is responsible for part of the design of a system), do you think it is OK for him to attend the walkthroughs of his subordinates? What problems is this likely to cause? What can be done to avoid these problems?

12 PERFORMANCE EVALUATION IN WALKTHROUGHS

American life is a powerful solvent. It seems to neutralize every intellectual element, however tough and alien it may be, and to fuse it in the native good will, complacency, thoughtlessness, and optimism.

George Santayana
Character and Opinion in the United States, 1920

Whenever managers discuss the concept of walkthroughs and team programming, several questions invariably come up: How can we evaluate the performance of the people who participate in the walkthroughs? How can the "good guys" be rewarded? How can the "dummies" be identified? As I have pointed out in previous chapters, managers are discouraged from attending the walkthrough, so they don't know how many bugs were found, how many idiotic programming techniques the producer used, or who thought of the brilliant way of solving the problem.

Surprisingly, many programmers and systems analysts feel the same way. "Why should I carry that clod?" they will argue. "The only reason his program works is that I found all the bugs! And my programs always pass the walkthrough the first time—with no bugs!" Naturally, these feelings are intimately connected with promotions, raises, and other forms of reward and recognition.

So, it's a problem that worries both managers and technicians. I will begin this chapter by examining the current situation in more detail; that is, how do managers evaluate the performance of their technicians in a non-walkthrough environment? I will then discuss the impact of walkthroughs on the process of performance evaluation.

12.1   Comments on Classical Performance Evaluation

In most of the organizations I have visited, as well as the ones I've worked in, programmers and systems analysts are reviewed once or twice a year.   The reviews range from informal chats with one's immediate supervisor to highly formalized rituals, complete with appropriate forms, and attended by two or three levels of management.   The result may be a promotion, a raise, or a pat on the back if things go well; or a rebuke, a demotion, or dismissal if things do not go well.

It's difficult to say whether the average performance review is fair, accurate, or honest. Also, it's difficult to make generalizations about the methods by which managers arrive at their decisions.   But there do seem to be some common characteristics to these reviews, and they are particularly interesting in the context of walkthroughs.

For example, I have noticed that *technical competence* is only a small factor in many performance reviews. When I was a junior programmer, I assumed that I would get a larger raise at the end of the year if my code was efficient, my programs well-documented, and my projects completed on schedule.   If my programs were three times more efficient than those developed by Charlie, the fellow who sat at the next desk, then naturally I would get three times larger a salary increase than Charlie. I soon learned that that wasn't necessarily so: I might get a larger raise if I had a better record of finishing my project on time, but raises and promotions had virtually nothing to do with the efficiency, maintainability, modularity, or overall *quality* of the product I was building.

Of course, conditions are more extreme in other organizations. In one large bank I visited in Montreal, the programmers and systems analysts spent quite a lot of time arguing about the technical virtues that were most likely to give them a large salary increase, until one programmer summed it up rather neatly by saying, "In the final analysis, it doesn't matter what kind of programs we write. If a program should take one month to develop, and we take two months, nobody notices. And if it should run in 512K bytes of memory, but we gobble up a megabyte or two, nobody cares— and if they *do* care, they'll just buy more memory."

"All they really care about," this programmer went on to say about her managers, "is that we show up for work on time, that we wear respectable clothes, that we work quietly at our desks, and— most important— that we not do anything to upset the users outside the MIS department."

Sounds pretty grim, doesn't it? Yet it is more accurate than many of us would like to admit. In this modern age, we rarely see the lone programmer toiling away by himself, single-handedly accomplishing technical miracles. It was true in the 1960s, and it is occasionally true in PC-oriented software development companies,[1] but it doesn't reflect the reality of modern life in the data processing department of most large organizations around the world.

Whether we like it or not, most MIS organizations are bureaucracies; they are part of what Apple Computer CEO John Sculley in [Sculley, 1987] calls "second-wave organizations." They consist of dozens, if not hundreds or even thousands, of people working on complex projects for other departments whose motivations and objectives they often don't understand. Modern MIS organizations are intimately involved with human bureaucracies, human systems, and human politics— and the people who excel in modern data processing careers are those who, *in addition to a requisite minimum knowledge of the technical details*, are talented at dealing with people.

In other words, if you are a programmer, it does matter whether you wear a clean shirt to work. If you are a systems analyst, it does matter whether you take a bath more than

---

1 Well-known examples in the PC world are Bill Atkinson, the author of MacPaint and HyperCard for the Macintosh, and Wayne Ratcliffe, author of dBASE-II.

once a week. It *does* matter whether you show up for work on time. Also, it does matter whether you can get along with your fellow programmers and systems analysts, as well as the manager and the users. In a majority of organizations, all of this behavior matters much more than the efficiency of your code or your familiarity with DB2, or OS/MVS, or the IBM 4341.

To the extent that technical competence *is* important, we have to ask how well a manager can judge the technical ability of his subordinates. Most managers will admit that this is an extremely difficult assessment to make, and that it may be impossible. Each programmer or systems analyst works on a different assignment, and each technician will argue that his project is more difficult than anyone else's, and that it may be the most difficult project the world has ever seen. Since there is not yet a universally accepted metric with which to measure the complexity of a computer system or the amount of functionality it delivers to the end user, we almost have to take the developer's word for it.[2] Of course, a manager with a great deal of experience may be able to use the following argument when evaluating the performance of one of his programmers, Fred:

1. Fred spent the last twelve months working on a general ledger system which barely works and which doesn't seem to completely satisfy the user's requirements.

2. When I was a junior programmer working on the IBM 360, I developed a general ledger program in one month, and it's been running without bugs since then.

3. Two or three other programmers who have worked for me during the past several years have managed

---

[2] This is changing, and we may have an industry in the mid-1990s that universally agrees on function points, McCabe's cyclomatic complexity metric, or the Halstead "volume" metrics for complexity. For more on this, see [DeMarco, 1982], [Halstead, 1977], and [Boehm, 1981], as well as an interesting review of complexity metrics [Curtis, Sheppard, and Milliman, 1979].

to write a general ledger program in two to six months without any major difficulties.

4.   Ergo: Fred must not be such a good programmer.

The trouble with all of this reasoning is that it's usually a false comparison, an attempt to compare apples and oranges. For example:

1.   It probably took the manager three months to develop his version of the general ledger program many years ago. Everyone, including Fred and the manager, would prefer to forget about those fuzzy periods of systems analysis and the tedium of testing, and remember instead the brilliant few days of coding.

2.   In the days of the IBM 360, the user probably didn't expect his general ledger program to do as much. He was likely to be far less critical of what he got; today, he is very fussy about the format and layout of screens and reports, and expects the general ledger program to interface with a dozen other existing systems, none of which has been properly documented.

3.   In the days of the IBM 360, the manager probably didn't have to document his program. Fred, on the other hand, has to fill out ten pages of forms, specifications, and status reports for every page of code he writes.[3]

4.   The IBM 360 worked. The modern-day Kludgevac 807 probably doesn't. Neither does the fancy relational, distributed database management system Fred is supposed to use.

---

[3] Capers Jones, whom many regard as the premiere statistician in the software engineering field, points out that every line of code written for the U.S. Defense Department is supported by 120-200 English words of documentation. For more amazing statistics like this, see [Jones, 1986].

5. On top of everything else, Fred is supposed to be using structured programming, structured design, and structured analysis, supported by a CASE product provided by a vendor that has just gone bankrupt. But since Fred doesn't know structured programming, design, or analysis, he's learning as he goes.

6. The boss conveniently forgets that Fred is maintaining three other systems while he works on his general ledger program.

7. Finally, the reason that the boss *is* the boss is that he *is* better than Fred—or at least he is able to convince everyone else that he's better.

I could continue, but the point should be clear. A manager can use his personal experience, his observations of other programmers and systems analysts, and his "gut" feelings to determine whether Fred has been doing a good job or not. But all he really can do is identify the high and low ends of the spectrum: the extremely good and the extremely bad technicians.[4] Ultimately, does it matter? In more and more organizations, workers expect (and receive!) nearly automatic salary increases, sometimes euphemistically called merit raises, as opposed to cost-of-living increases, regardless of whether their performance was 10 percent above or below average. Only the people at the high and low ends of the spectrum merit special attention. The extremely talented technician is given a 20 percent salary increase instead of the usual 10 percent—even if his performance was ten times above average. The extremely incompetent technician is

(a) given a 5 percent salary increase—after all, we don't want to hurt his feelings; or

(b) given *no* salary increase—maybe he'll take the hint and quit in another year or two; or

---

[4] This assumes, of course, that the manager *does* have some technical knowledge and experience. Some managers don't; for example, they may be ex-accountants or shoe salesmen who have been asked to take over the MIS department.

(c) transferred to the (shudder!) maintenance department to die a slow death; or

(d) fired—an option that is used, quite literally, as the last resort in most MIS organizations.

12.2   The Impact of Walkthroughs on Performance Evaluation

The main point of the preceding discussion was to put the subject of performance evaluation in some perspective. It is appropriate to ask how technicians are evaluated in a conventional organization, and then to ask how walkthroughs will impact that process.

Let's begin with a simple observation. Since a technician's performance is largely based on non-technical criteria, walkthroughs should have no impact on the evaluation process. Management still can determine whether their programmers show up for work on time, whether their systems analysts interact well with the user, and whether members of the staff take a bath at regular intervals.

But is it as simple as that? The walkthrough approach *does* introduce a major change in the social dynamics of the MIS organization: The programmer or analyst is expected to attend walkthroughs, participate in the review of other people's programs, and accept constructive criticism of his own programs. Thus, the manager may wish to include the following questions in his review of the technician:

1.   Did he make himself available for walkthroughs or did he refuse to participate? Did he insist on being a loner, and if so, did he produce enough to justify such anti-social behavior?

2.   If he accepted an invitation to a walkthrough, could he be counted on to attend—or did some last-minute crisis frequently prevent his attendance?

3.   Did he invest the necessary time and energy to study the programs and designs of other technicians in the organization, or did he do a superficial job of preparation?

4.   Did he generally follow the procedures for walkthroughs described in Chapters 5, 6, and 7?

5.   Was he able to give and receive constructive criticism easily, and did he concentrate on critiquing the product rather than the producer?

Many of these questions will be difficult for the manager to answer, simply because he is not present in the walkthrough and thus cannot judge firsthand what the technician is doing. In any case, it does seem that the walkthrough approach may "hide" certain information from the manager regarding the programmer's technical competence. However, just because the walkthrough takes place behind a closed door does not necessarily mean that it functions as a black box. In the organizations where I have worked, the manager was able to identify the people at the high end and low end of the spectrum. The manager could *not* easily tell which of his average people were better than the other average people—but he probably couldn't tell before walkthroughs were introduced, either!

Why is it that the manager can identify the very good and the very bad people? Part of it has to do with the range of performance, whether one is measuring productivity or quality of the work performed. While walkthroughs generally improve *everyone's* productivity and the quality of *everyone's* product, it is still true that the talented programmers will produce superior products.[5] In addition, the office gossip will make it very clear who the talented technicians are and who the incompetents are. If Charlie requires seventeen walkthroughs before his program is approved, the manager

---

[5] A classical study indicates that some programmers can finish their projects 28 times faster than others, and their programs would be 10 times more efficient than the programs written by others. See [Sackman, Erickson, and Grant, 1968] for more details on this.

will certainly hear about it; indeed, the other programmers and analysts may refuse to participate in the walkthroughs if they feel they have to carry Charlie.[6]

At the same time, the manager will hear about the walkthrough in which Mary's program is praised by everyone for its documentation, its elegance, and its efficiency.   Hence, the hero who single-handedly saves the project with his brilliant work won't be made invisible because of the walkthrough; and the dummy will find that the walkthrough doesn't hide his incompetence from the rest of the world.

As for the others, the manager has to accept the fact that he will lose some information on which to judge the technical competence of his people.   But things average out. If Fred finds two bugs in Mary's program on Monday, there's a good chance that Mary will find two or three bugs in Fred's program on Tuesday.   If that's the case, should the manager care whether Fred had more bugs than Mary?

The manager can always ask to look at Fred's program outside the walkthrough.   Fred's program will reflect some of the suggestions, improvements, and criticisms of the various reviewers, but it will probably still show a lot of Fred's personal handicraft.

There is one last comment to make about the impact of walkthroughs: Sometimes they provide *more* information to the manager than the classical approach and thereby help deal with a phenomenon that my colleague Tim Lister calls the "living legend."   Consider the case of Veteran Vic, who has been with the organization for twenty years.   Vic played a major role in developing the financial reporting system, which runs six hours a day, and consumes 12 megabytes of memory and 43 spindles of disk storage.   During the first two years of

---

6 Some people have suggested that the programmer's performance could be judged by the number of walkthroughs required to produce an approved product.   This criterion may be OK for extreme cases (like seventeen walkthoughs), but is probably dangerous if the manager tries to establish a threshold of two or three walkthroughs as a measure of competence.   If the reviewers know that this is Charlie's third walkthrough, they may be more inclined to approve the product, even though it is still inferior; and naturally Charlie will not be in a mood to accept constructive criticism.

production, when the system was extremely shaky, Vic spent many long nights in the computer room, fixing bugs, rerunning jobs, writing "quick-and-dirty" programs to restore clobbered master files, and the like. Having been around so long, Vic has now acquired the title of Senior Software Systems Consultant; he's a living legend, and the other programmers and analysts treat him with the deference that the title requires.

But has anyone seen any of Vic's code for the past few years? Does anyone know whether Vic *can* code? In fact, does anyone know whether the financial system was developed *because* of Vic, or *in spite of* Vic? Many of these questions can be answered if Vic is requested to subject his programs to a walkthrough. In some cases, Veteran Vic will truly be recognized as a veteran— and the junior programmers and analysts will have an opportunity to learn from a master of the trade. In other cases, Vic will have a hard time maintaining his status, his title, or even his job. I have seen one or two veteran programmers resign— with a great deal of righteous indignation— rather than be subjected to the indignities of a walkthrough.

To summarize, the manager may find that he loses a little information because of the "secrecy" of a walkthrough; however, it opens up the systems development process and exposes the living legend who might otherwise have escaped detection. Walkthroughs do *not* destroy the manager's ability to detect extremely talented and extremely untalented people; nor do they interfere with the manager's ability to judge non-technical aspects of the technician's performance. And finally, it should be obvious at this point that the criticisms of a product and the errors found in that product during a walkthrough should *not* play a part in the producer's performance evaluation.

QUESTIONS FOR REVIEW AND DISCUSSION

1. How often do performance reviews take place in your organization? Does a performance review always take place at the same time as a salary review? Is it the

general opinion of the programmers and analysts in your organization that the review is fair?

2. What are the major factors in the evaluation of a technician in your organization? How many of these factors are technical, and how many are non-technical?

3. How accurately do you think a manager can judge the technical competence of his subordinates? By what methods does he rank their technical abilities?

4. Does your organization provide different salary increases for *every* technician, or does it place each technician into one of a few categories— such as good, bad, and average? If the technician is placed into categories, is it generally true that everyone who falls into the same category gets the same percentage salary increase?

5. If walkthroughs are introduced into your organization, will a technician's performance be judged partly on the basis of his willingness and ability to participate in walkthroughs?

6. Do you think that a manager will be able to identify very talented people and very "untalented" people in a walkthrough environment?

7. Do you have any living legends in your organization? Will walkthroughs help expose their talents, or lack of talents, for everyone to see? Why or why not?

13 BUILDING TEAMS

> We pass the word around; we ponder how the case is put by
> different people; we read the poetry; we meditate over the
> literature; we play the music; we change our minds; we
> reach an understanding. Society evolves this way, not by
> shouting each other down, but by the unique capacity of
> unique, individual human beings to comprehend each
> other.
>
> Lewis Thomas
> *The Medusa and the Snail. On Committees,* 1979

In Chapter 3, I introduced the concept of programming teams. I pointed out that many organizations are unwilling or unable to form true programming teams, but that they can implement walkthroughs, even if they have a conventional project organization.

Regardless of whether or not your MIS organization plans to organize true programming teams, any attempt to implement walkthroughs necessarily involves some teamwork. After all, everyone in the walkthrough has to agree to a certain set of rules and procedures, and everyone has to agree that their purpose is to review the product rather than the producer.

But is it likely that your organization will develop the type of programming teams described in Chapter 3? Is it likely that MIS management will be willing to assign a project to a team and then let them manage their day-to-day activities completely? As you might imagine, the concept seems radical to many organizations— almost communistic. In such organizations, the idea of programming teams is either entirely abandoned or the teams function in an underground sense. That is, the programmers and systems analysts know that they are working as a team, and perhaps even the first level of management knows, but all of the members of the team continue carrying out their normal day-to-day activities so that the "normal" management hierarchy will not think that anything has changed.

I will not discuss strategies of underground revolutions in this chapter, nor will I discuss organizational guerrilla warfare; if you are interested, I suggest that you read [Townsend, 1984] or [DeMarco and Lister, 1987]. Instead, I will assume that you, as a manager, are interested in forming programming teams within the normal framework of your organization. Since this book is not a management textbook per se, my discussion will be limited to a series of suggestions for successful formation of programming teams within your organization.

13.1   Guidelines for Forming Teams

The main ingredient in forming programming teams is a commitment by the manager: While underground teams *can* form spontaneously among the programmers and systems analysts, a host of political problems make it difficult for them to continue. So, to begin with, you must *want* programming teams to form and flourish; if you are ambivalent, it's probably better not to bother with the idea.

The second main ingredient, though it's not absolutely necessary, is a project for the team to work on. Once the team concept catches on, you will find that it doesn't have to be project-oriented. In fact, a team can be working on three or four small projects simultaneously. But in the beginning, a single project gives everyone something to focus on. Thus, wait until you have a development project of about the right size and then assign it to the team.

With these two basic ground rules, most MIS organizations simply form a team, assign them a project, and then deal with unusual situations as they arise. In most cases, the team begins to show a personality of its own after a few months, and both management and team members will learn how to deal with the new environment. In most organizations, neither the programmers and the systems analysts nor the managers can claim to be expert psychologists, so everyone has to blunder along with their amateur understanding of the social dynamics—but they DO manage to get along.

With this in mind, you may find the following suggestions helpful in building programming teams:

1. *Let the team evolve naturally.* Many organizations form their programming teams by taking a group of people who already were assigned to work on a project and telling them that they are now part of a new social organization. Only later does management realize that two of the team members violently dislike each other and that two others argue incessantly about abstract programming issues. By the time this friction is discovered, the team may be on the verge of collapse. The moral: You can't create teams by an arbitrary decree. The fact that five or six people have been working together on the same project is some indication that they can communicate with one another; however, you should talk privately with each individual to see whether personality problems or other issues may prevent them from functioning as a true team.

2. *Don't disband the team at the end of the project.* As I have mentioned, many organizations form their first teams around a specific development project. At the end of the project, it is natural for management to reassign the programmer/analysts to other projects, thus effectively destroying the team. Since it is a rather delicate matter determining which technicians can work well together, and since it usually takes a few months for even the best of friends to learn how to work together in a team, destroying the team at the end

of the project wastes that investment of time and energy. Let the team stay together and assign it a new project.

3. *Encourage fresh blood in the team.* The major flaw with the above suggestion is that a group of programmers and analysts may become technically stale after working together for a few years. They get themselves into a rut and keep using the same kind of approach for every new project, even when the situation calls for fresh, new approaches. In addition, people often become bored with one another after working together for years. For these reasons, it's healthy to allow and even encourage some turnover of people in the team. This change will usually happen without overt actions from management: One team member will leave to take a better job in another company, another may leave because his family is moving to East Podunk, and so forth. The new programmer or systems analyst who joins the team brings his own experiences and perspective, which will be assimilated into the collective experiences of the team. As Gerald Weinberg says, "Programmers come and go; the team abides."

4. *Respect and trust the team.* Without trust and respect, it is impossible for the team to function. Nevertheless, many MIS managers display mild distrust of the team's day-to-day activities and begin dabbling in its affairs. It is perfectly proper for management to ask the team to establish periodic checkpoints and milestones, e.g., on a weekly or monthly basis, depending on the size and nature of the project, and to ask for a full accounting of progress at each checkpoint meeting. But if it looks as if the project is getting behind schedule, or if it appears that two team members are spending a lot of time arguing with each other, the manager should resist his first impulse to step in and solve things himself. If necessary, the entire team can be gathered together and confronted with manage-

ment's perception of the problem; such occasional showdown meetings allow the team to maintain its own integrity and still give management the satisfaction of seeing the problem openly aired and resolved.

5. *Reward and punish the team equally.* In some cases, the team members themselves may suggest to management that one of their members has done an outstanding job and that he deserves special recognition. This commendation rarely occurs, though, and the manager is courting disaster if he tries to make his own decision about the relative contributions of each team member. As I pointed out in Chapter 12, performance evaluation is often based on non-technical factors; thus, it is possible that a manager could decide that team member A deserves a larger salary increase than team member B without ever having to decide whether A's programs are better than B's programs. However, this individual evaluation is likely to introduce a destructive element of competition among the team members, making it more difficult for them to function effectively as a team. So, unless it is absolutely impossible, I advise that you reward or punish the team members equally.

6. *Protect the team from outside pressures.* In a typical organization, a programming team will be an alien unit within the bureaucracy. One of the most useful things that the manager can do is to protect his team members from the meetings, the memos, the Muzak, the forms, and the administrative pressures which would otherwise consume their time, energy, and patience. Often, this can be accomplished by moving the entire team into a separate working area, far removed from the rest of the Mongolian horde; this creates an air of excitement and mystery for the team, even if their physical working conditions are less glamorous and comfortable than those enjoyed by the rest of the MIS organization. This "skunk works" approach

has been widely used in various industries; I can testify from personal experience that it works in software development groups!

QUESTIONS FOR REVIEW AND DISCUSSION

1. If programming teams are not officially endorsed by your organization, do you think that they could function on a sub rosa basis? Could the programmers and analysts themselves form such a team, or would first-level managers need to participate actively?

2. How can you tell whether a group of programmers and systems analysts are likely to be able to work together as a team? Will *they* know? What kind of questions can you ask them to determine whether the individuals are compatible?

3. In your organization, are programmers and systems analysts normally reassigned at the end of a development project? Is it possible for them to continue working together on a new project?

4. Do you feel that members of a programming team should be rewarded or punished equally? Why or why not?

14 THE FUTURE OF WALKTHROUGHS

No one understands the kind of creative frenzy that causes
people like Bill Atkinson not merely to imagine HyperCard,
but to make it real; or Dan Bricklin and Robert Frankston to
dream up Visicalc and bring it to life. But we can all agree
that it's worth nearly anything to have people like them
doing what they do.

Jerry Pournelle
"Reuniting the Two Main Streams of the Micro Revolution:
It Started in 1987," *InfoWorld,* December 28,1987

Walkthroughs have probably existed since the first
computer program was written; indeed, they were heavily
(though informally) used all through the 1950s and 1960s,
because computer hardware was so scarce and expensive that
there was enormous pressure to get the program working
correctly the first time. Walkthroughs have been discussed in
computing literature since at least 1971, when Gerald
Weinberg's popular *The Psychology of Computer Programming*
[Weinberg, 1971] was published; and they have been
mentioned frequently in books and articles about software
engineering and structured systems development techniques.
Indeed, they have been practiced without significant changes
during the 1970s and 1980s, and it would be easy to assume
that they will continue to be practiced in large MIS
organizations through the 1990s as well.

But I am convinced that we will see major changes in
the way walkthroughs take place during the next few years—

changes that will be at least as significant as the changes I have witnessed since the first edition of this book was published in 1977. The purpose of this chapter is to explore some of these future possibilities. While my crystal ball is not guaranteed to be any more accurate than yours, I hope that my visions of the future will at least make you start doing some serious thinking about *your* visions of the future.

14.1 Walkthroughs Will Become More Specialized

In my work as a consultant, I have often found myself thrust into a walkthrough of a system I knew nothing about, working with system developers I barely knew.[1] Worse, the walkthrough involved a specialized area of an application about which I was hopelessly ignorant: stock transfers, missile guidance, or molecular analysis of organic compounds. Nevertheless, I was usually able to avoid making an utter bozo of myself by concentrating on the *completeness* and *integrity* issues discussed in Appendices A, B, and C.[2] Are the dataflow diagrams balanced? Are the interfaces properly documented? Do the data elements on the structure chart or dataflow diagram match the definitions in the data dictionary? All of these questions could be asked without any knowledge at all of the underlying application.

Today, the growing sophistication of automated tools (e.g., CASE tools) means that much of this level of error-checking can be done automatically; indeed, it may be carried out by the developer himself before the walkthrough even takes place. On the other hand, as I have mentioned throughout this book, such automated tools are showing up

---

1 How did this happen, you ask? Usually because an MIS manager, having listened to me extol the virtues of walkthroughs, would say to me (with a sly, crafty grin on his face), "Well, it just so happens that there's a walkthrough taking place right now: The inventory control project team is reviewing the dataflow diagrams from their interviews with the users. How would you like to sit in? It'll give us a chance to see if this stuff really works!" You can imagine how thrilled the project team is with this kind of intrusion...

2 In other cases, I avoided looking like an idiot by the simple expedient of keeping my mouth shut. And then, of course, there were a few times when I opened my mouth and *did* sound like an idiot.

more slowly in MIS organizations than one might think by reading articles in *Datamation* or *Computerworld*; while leading-edge organizations have had many such tools since the early or mid-1980s, it will probably be the mid-1990s before automated CASE tools are as common and widespread as, say, on-line, "dumb" terminals are in today's systems development environment. So, for the next few years, "ignorant" outsiders like me can still make a contribution in a walkthrough.

But when the automated tools are sufficiently widespread, we will see more and more specialization in the walkthroughs. The most obvious specialization is the application itself: A walkthrough of an airline reservation system will require participants familiar with the ins and outs of airline reservations. And there will be specialization in terms of the overall industry or working environment in which the company operates; a good example of this is the U.S. Defense Department's DOD-2167A standard for developing defense systems.

Another area of specialization will involve the "human interface" of information systems: With increasingly sophisticated terminals, workstations, voice recognition systems, OCR, "mouse" input devices, etc., the programmer whose only expertise is COBOL will have an increasingly difficult time participating in a walkthrough of the "user implementation model" of a system. We will need instead (or in addition to) people who are human interface experts. Similarly, MIS organizations building expert systems will probably need experts at "knowledge acquisition" (see [Keller, 1987] for more discussion of this) to ensure that their system's knowledge base is truly representative of the expertise of the human experts.

What does all of this mean to the manager committed to walkthroughs in his MIS organization? My primary recommendation is that you should not consider the list of participants discussed in Chapter 4 to be "frozen." As automated tools eliminate more and more of the trivial system development errors, be prepared to add specialists who can help eliminate more and more of the difficult but subtle errors.

14.2   The Economics of Walkthroughs Will Be Explored More Fully

As I pointed out in Chapter 10, studies by Barry Boehm, Capers Jones, Glenford Myers, and others have indicated that walkthroughs are usually successful in finding as many as 80 percent of the errors in a program, and that they are often more cost-effective than other methods of error detection and removal.   But the hard, quantitative evidence is still rather scarce; as a result, some MIS managers are still cautious about investing too much time and resources in walkthroughs.

But the "metrics" movement is gradually gathering energy throughout the entire systems development profession, as more and more organizations realize that unless they establish a company-wide program for software metrics, they will probably never control MIS costs, schedules, or quality.   (An excellent discussion of this trend can be found in *Software Metrics* by Robert Grady and Deborah Caswell [Grady and Caswell, 1987].)   And one of the results of this effort to develop company-wide metrics will be much more detailed and thorough measures of the effectiveness of walkthroughs.

14.3   The Boundary Between Product Development and Product Walkthrough Will Blur

An obvious assumption throughout this book is that there is a very sharp line between the *development* of a product (whether code, design, or specification) and the *review* of that product; as I pointed out in Chapter 5, for example, it is incumbent upon the developer to give all of the walkthrough participants at least two days' advance notice that he is ready for a walkthrough ... and he can only do this when he is finished, or so nearly finished that he has appropriate materials to distribute.

This may continue to be the way things are done in many large MIS development projects.   But there is another trend that is becoming important, and which will have an important impact on walkthroughs: the *rapid prototyping* approach to systems development. Known by various names— rapid systems development, joint applications development (or JAD), evolutionary prototyping, etc.—this approach typically brings users and systems developers into close contact for a short, intensive period of time, at the end of which a prototype, functionally complete system has been built. An excellent discussion of this approach can be found in *Application Prototyping* [Boar, 1984] and *Rapid Prototyping* [Connell and Shafer, 1988].

What will this do to walkthroughs? For one thing, it will mean that most of the walkthroughs will concern *actual* (prototype) systems instead of paper models of a system. Most advocates of rapid prototyping agree that paper models (dataflow diagrams, etc.) are still important at the very beginning of the project, in order to provide an overview of the system and to partition it into major subsystems—but the bulk of the work is done with real input screens and real output displays on real terminals or workstations.

More importantly, we will see development and review taking place at almost the same time, as the prototyper and the user work together to build, modify, and evolve the system that the user wants. By thus virtually eliminating the "turnaround" problem that I have alluded to in previous chapters, we should see a dramatic improvement in the productivity of both the development and the review processes.

I am very much in favor of this scenario; and I think it will become a more and more common systems development scenario, regardless of whether I or other MIS managers approved of it or not. My only concern is that this rapid prototyping environment, with its "instant feedback" approach, may preclude the quiet, thoughtful contemplation that characterizes the systems development environment in today's MIS organization. For better or worse, the current environment offers the developer multiple opportunities, both before and after a walkthrough, to sit back, stare at the ceiling, and *think* about the system he is building ... to think, for example, about error situations that the user himself might not have thought of in the heat of the moment of a prototyping session.

There is no reason, of course, why thoughtful contemplation must necessarily disappear in a rapid prototyping environment; it is more a question of whether the prototyping tools are used and managed properly. We had a similar situation in the 1960s and 1970s: Many MIS organizations found that when programmers were first given on-line time-sharing terminals with which to enter their source programs, they began composing programs extemporaneously at the terminal, rather than thinking carefully about the logic they wanted to construct. I am concerned that some MIS organizations will have a similar problem, on a much grander scale, when they first use a prototyping approach.

14.4 High-Tech Support Tools Will Facilitate the Development of Groupware

Rapid prototyping, discussed above, is only one example of "high-tech" tools that are changing the way we build information systems. There are many other "productivity tools" that are now widely used: interactive debuggers, source code librarian packages, and dozens of other examples. But most of these tools are aimed at making an individual programmer/analyst more productive at the work that he does by himself.

And yet, with the exception of the "cowboy programmer" paradigm discussed in Section 14.5, most information systems are built by groups of people; indeed, the very large systems (those with which we have the most difficult problems of productivity and quality) are built by groups of companies, which oversee groups of divisions or departments, which consist of groups of groups, which contain groups of individuals. To make matters worse, these groups are often not located in the same site: It is easy to put five people in a single room, but it is much more difficult to put 500 or 5,000 in a single room—especially if they work for a dozen different organizations.

This geographical separation of the workforce also complicates the walkthrough process. If a portion of a system is being developed by someone in Houston, but interfaces with other portions built by groups in New York, London, and San Francisco, how can they effectively communicate with one another in a walkthrough? Bringing all of these people together in a single conference room for a 30-minute walkthrough is hardly practicable!

We are beginning to see the hardware and software support tools that will allow us to deal with this situation. They include the following:

- Local area networks and other high-speed networks that support electronic mail, file servers, etc.

- The concept of "project dictionaries" that contain all of the relevant information about an ongoing systems development project— all of the diagrams, data dictionaries, source code, schedules, budgets, test data, etc.

- Network versions of CASE tools. These are already available from several CASE vendors, and will become universally available by the early 1990s.

- Electronic "post-it" notes that allow a reviewer to attach notes and comments to any portion of an author's document without altering the document itself. Such facilities exist today for text documents (e.g., the "For Comment" program for IBM PC document files); it is reasonable to assume that we will soon see similar facilities attaching comments to dataflow diagrams and other graphical models.

- Utility programs that will allow remote workstations to be locked into complete synchronization, so that individual developers can work on the identical display simultaneously. One example of such a program is *In Synch*, available from American Video Teleconferencing Corp. in Farmingdale, NY. Another example is *The Coordinator*, available from Action Technologies in Emeryville, CA.

- Programs for gathering and evaluating group decisions— so that several remotely located walkthrough participants can review the same product and reach a decision in the manner suggested in Chapter 6 of this book. An early version of such a program is *Q System 100*, available from Reactive Systems, Inc. in Englewood, NJ.

- Teleconferencing facilities, so that individuals can have face to face group meetings even though they are in separate geographical locations.

It is reasonable to assume that the hardware and software technology will continue to advance for the next several years, and that we will have more and more powerful tools available. But as we know from watching the introduction of other new technologies, the initial reaction of most MIS organizations will be to use this new technology to do the same thing they are doing now, only faster and cheaper and better.[3] That is, they will continue to follow the current paradigm of letting an individual do his own individual work, after which it will be reviewed by other individuals.

The real paradigm shift will come when MIS organizations undergo a major change in their culture: Instead of individually developed programs or modules, we will begin to see jointly developed programs. The analogy that we could draw for the pre-network organization would be that of several programmer/analysts with their hands all on the same pencil as a line of code is being written, or as a bubble is being drawn

---

[3] I first heard this described by my colleague Rob Thomsett as "Fubini's Law." Some examples: When Alexander Graham Bell first presented the telephone to dubious audiences, he suggested that it could be used as a marvelous way of listening, at home, to *live* concerts—i.e., as a more efficient one-way communications device. And when the motion picture camera was first introduced, it was seen by many as a marvelous way of recording live plays, so that people would not have to sit in a theater to watch Shakespeare. For more on this, see *Nations at Risk* [Yourdon, 1986].

on a dataflow diagram. The electronic equivalent is, of course, much more efficient.

Can such a thing happen? The issue here is not technology, but technology transfer. And the question is whether a group of loosely connected individuals (which is the context in which most walkthroughs take place today) will be willing to relinquish their individual pride of ownership in the products they build, in order to *truly* work together as a programming team, in the sense described in Chapter 3 of this book.

I would be the last to suggest that this will become a universal phenomenon, but I think it *will* happen in some organizations. And once such a team has had the opportunity to work in this fashion, they will never again accept any other mode of working. On the other hand, there will still be room for the current kind of system development organizations ... and there will also be a growth in what I describe below as the "cowboy programmer" paradigm.

14.5   The "Cowboy Programmer" Paradigm Will Develop Its Own Walkthrough Techniques

This book has taken the position that one-person projects are (a) an anomaly, and (b) simple enough so that walkthroughs are useful, but not critical. This is probably a fairly accurate reflection of the "real world" in most large MIS organizations where 10-person projects are considered small, where 100-person projects are "normal," and only the 1,000-person projects are considered large.

But there is another world that is having an increasingly important role in large organizations: the system development projects for PCs. Even in this world, the projects are becoming much, much larger than they were in the late 1970s, when it was common for a single programmer to create a spreadsheet product like 1-2-3, or a database management package like dBase-II. More and more PC-based information systems are just as large—and thus just as complex—as the minicomputer and mainframe systems of the 1960s and 1970s.

But with increasingly powerful PC hardware (of which the 80386-class machines and Motorola 68020 and 68030 engines are the best examples as this book was prepared in early 1988) and increasingly powerful PC development tools, it will still be possible for individuals to build dazzling, awe-inspiring new systems. Apple Computer's HyperCard, developed primarily by Bill Atkinson (who achieved earlier fame as the developer of MacPaint for the Macintosh), is probably the premiere example of such single-handed wizardry in the late 1980s.

And it is an undeniable fact that our American culture glorifies such lone cowboys. Such glorification does not deny the existence of large, Mongolian-horde projects, but it inspires each new generation of college computer science students— or even high-school-age hackers— to dream up the next generation of PC programs.

But there is an ironic twist here. Even though such programs can be, and are, conceived in the minds of one or two people, they succeed only if they impress and dazzle *and are used by* a large number of people; it is interesting, for example, to note that Apple Computer began bundling HyperCard *free* with all sales of Macintosh computers when the software package was released in the summer of 1987. And if such a product is going to be used by hundreds of thousands of people around the world, it has to work. It can have some rough edges, but it can't be riddled with bugs— it can't destroy users' files or abort unexpectedly.

Thus, even the cowboy programmers are realizing that they need some mechanism for assuring that their products are of high quality. Listen to what Bill Atkinson had to say about the development of HyperCard in [Goodman, 1987]:

> What I'm finding with HyperCard is that this trend in the way I work is progressing. I'm working more with other people. There are twenty people who've got their fingers in the pie on HyperCard. There are four that are writing serious chunks of the code. It's still basically my baby. I insist that everything be right, and I read all the code. I probably have written 70 percent of the code that's in there. Dan Winkler is the next major contributor to the code. I needed somebody who could keep up with me, who would go with my style. Dan was

> willing to learn from me and pick up my style. We work back and forth, we hand code back and forth. We both work here at my house, and I see, in general, I'm able to accomplish a lot more by working with other people.

This kind of development will, I think, be the most exciting paradigm of the 1990s for those of us who still have the privilege and the responsibility of building high-quality information systems.

APPENDIX A:

GUIDELINES FOR ANALYSIS WALKTHROUGHS

$\mathbf{T}$his appendix provides specific guidelines for the conduct of systems analysis walkthroughs—walkthroughs of products whose primary purpose is to describe the user's *requirements* for a new system. It assumes that the project team is using the kind of structured analysis techniques espoused in such books as [Yourdon, 1989], [Ward, 1984], or [DeMarco, 1978]. However, there are various other structured analysis methodologies (see, for example, [Martin and McClure, 1985], [Orr, 1977], [Jackson, 1983], [Higgins, 1979], [Hansen, 1984], or [Gane and Sarson, 1977]), and the guidelines below may have to be modified slightly for them.

Reviewers should check for four major criteria in a structured specification: *completeness, integrity, correctness,* and *quality.* Each of these is discussed in more detail below.

A.1   Completeness

It is impossible to conduct a meaningful walkthrough of an incomplete specification. Yet I see this happening time after time in many MIS organizations. It is quite proper, of course, to review a complete, consistent *subset* of the overall specification of user requirements, but if that subset is missing important information, the reviewers cannot assess its quality.

Thus, the very first questions that the reviewers should ask are as follows:

- Are the dataflow diagrams complete? Are all of the "bubbles" numbered and named? Are all of the data stores, terminators, and dataflows labelled? Is the overall diagram numbered or labelled so that you can tell what subset of the overall specification it represents?

- Are the data dictionary definitions complete? (It would amaze you to see how many analysis walkthroughs take place with no data dictionary!) Have all of the data elements been defined? Have the components of each data element definition been defined, too? That is, if the data dictionary contains the notation

$$A = B + C$$

have B and C been defined?

- Has the analysis team put too much information in the dictionary, adding to the problem of maintaining it when changes inevitably occur?

- If any of the bubbles in the dataflow diagram are primitives, is there an associated process specification, or minispec, for that bubble? Is it complete?

- Is there an entity-relationship diagram that describes the data objects in the system and the required relationships between the objects in this system? (Systems analysts who became acquainted with structured analysis in the late 1970s by reading such books as [DeMarco, 1978] and [Gane and Sarson, 1977] may not be familiar with this notion. More recent structured analysis books such as [Ward, 1984], [McMenamin and Palmer, 1984], and [Yourdon, 1989] cover this topic thoroughly; also, many books on database and information modeling, such as [Flavin, 1981], [Howe, 1983], and [Inmon, 1988], describe entity-relationship diagrams in detail.)

- If the system being specified is a real-time system (for example, an embedded system, a process control system, a telephone switching system, etc.), is there an associated state transition diagram for each control process in the dataflow diagram? Analysts working on conventional (non-real-time) systems can ignore this modeling tool, though the implementers of such systems may need them when developing their design of the system.

Many aspects of this completeness check are almost clerical and could be accomplished by the walkthrough coordinator or by an administrative person assigned to the project. The team could agree that an analysis walkthrough won't even be scheduled until the completeness criteria described above have been met. (With an appropriate CASE tool, most or all of this could be accomplished by computer, simplifying the job even further.)

A.2   Integrity

After ensuring that the specification is complete, the reviewers need to examine the integrity of the specification documents, ensuring that they do not contain inconsistencies or contradictions.

Like the error-checking described above, much of the integrity testing is fairly mechanical. It is perhaps too complex to be given to a secretary or administrative person (though many would disagree with that assessment), but it certainly *could* be done by a computer. Indeed, many of the analyst workbench CASE tools described in earlier chapters of this book derive their greatest value from their completeness testing and integrity testing capabilities.

Until the mid-1990s, though, the majority of MIS organizations probably will have to depend on manual error-checking to ensure the integrity of the specification. Thus, the reviewers in the analysis walkthrough should look for errors and problems of the following kinds:

- *Undefined data elements*— a dataflow diagram may show the existence of a data element (i.e., an input to some process or an output from a process) that is not defined in the data dictionary. The worst and most blatant case is where the data element is not defined at all; a more subtle error is where the dataflow diagram refers to a data element that is similar to, but not exactly the same as, an entry in the data dictionary. It's best to correct such minor errors before they escalate into major misunderstandings.

- *Undefined data stores*— each data store on the dataflow diagram should be defined in detail in the data dictionary. The blatant errors are the ones in which the dataflow diagram shows un-named stores, or stores with no definition at all in the data dictionary. The more subtle error is the DFD whose data stores have names that are similar to, but not exactly the same as, entries in the data dictionary.

- *Undefined references to a data element within a process specification*— the process specification describes the business policy for one small, well-defined portion of the system. It should be written in an acceptable form (with decision tables, structured English, pseudocode or some other acceptable form), and the references that it makes to various data elements should be examined to confirm that the data elements are legitimate entries in the data dictionary or local terms within the process specification.

- *Process specifications whose inputs and outputs do not correspond to the dataflow diagram*— the bottom-level, or "atomic" bubbles in a dataflow diagram show explicitly, in a graphical form, what inputs are required by the bubble and what outputs are produced by the bubble. Those should correspond to the inputs and outputs that are described in that bubble's process specification.

- *Balancing errors*— the concept of balancing is described thoroughly in virtually all books on structured analysis: The inputs and outputs associated with a bubble in a "parent" dataflow diagram must be the same as the net inputs and outputs for all of that bubble's associated "child" dataflow diagrams. This almost always requires the reviewers to consult the parent diagram, the child diagrams, *and* the data dictionary. The parent diagram may have a bubble #2, for example, with an input of "customer-name"; for proper balancing, there should be a related "child" dataflow diagram labelled as "Figure 2" (corresponding to the parent bubble #2), and the total net inputs to Figure 2 should be equivalent to "customer-name". Note that the child diagram, Figure 2, may show inputs of "first-name" and "last-name" and might *not* show an input of "customer-name". Only in the data dictionary might we find that "customer-name" is defined to consist of "first-name" and "last-name"; if this were the case, then the parent and child diagrams would be in proper balance.

- *Infinite sinks*— or "black holes," as some analysts call them in a comparison to the dense stars whose gravitational field is so strong that not even light can escape them. There are two common examples of an infinite sink: a dataflow diagram that has one or more inputs, but no outputs (what good is a system that gobbles up information, but never produces any output?), and "write-only" stores that have one or more *incoming* dataflows, but no outgoing dataflows. The second case may be a little more subtle than it appears: It is possible that the data store is created by the system for eventual use by another system, or by one of the terminators outside the system (consider, for example, a system whose purpose is to create market research data that can be accessed directly by customers outside the system). In this latter case, the data store should be shown as *being outside the system boundary* in the context diagram (the diagram in which the entire system is shown as a single bubble, and which shows the inputs and outputs between the system and its environment).

- *Spontaneous creation*— the most common example of this is a bubble that has output flows, but no input flows. With the exception of a random number generator, the analysts should ask how such a bubble is able to spontaneously (without any information or inputs from the outside world) decide when and what kind of output to produce. A related example is a read-only store: a store from which data elements are read, but which is never updated or written to.

- *Errors involving conservation of data*— I am indebted to my colleague Chris Gane [Gane and Sarson, 1977] for this pithy description of a common error in structured analysis models: "What comes out must have gone in." Quite simply, it means that the reviewers must examine each bubble in the dataflow diagram to convince themselves that the outputs produced by the bubble are, in fact, derivable from the inputs. If a bubble shows lead as its only input and gold as its only output, something is wrong.

- *Superfluous inputs to a bubble*— each bubble in a dataflow diagram should receive only those inputs that are essential for the performance of its task. This seemingly obvious concept is violated by many systems analysts who are accustomed to second-generation batch computer systems where an entire record would be passed from program to program— with each program extracting those fields of data that it needs. Today's programmers and systems analysts usually agree that this approach should be avoided if they have a database management system that can retrieve individual fields of data from a data store. In any case, there is no excuse for sending large amounts of unnecessary data elements into a bubble when developing the essential model of user requirements.

A.3   Correctness

Once the reviewers are convinced that the specification is complete and internally consistent, then they can move on to the heart of the matter: Does it truly describe the user's requirements for a new system? This is a difficult job in most cases because the systems analysts usually are not experts in the user's business (a reason for including one or more users in such walkthroughs!).

Systems analysts now distinguish between three different system models: the essential model, the user implementation model, and the systems implementation model to distinguish between "two" requirements and two aspects of the "solution" to the system: that part imposed by the user himself (the user implementation model), and that part invented by the systems developers (the systems implementation model) as the best use of current hardware/software technology. (See [Ward, 1984], [Ward and Mellor, 1985], [McMenamin and Palmer, 1984], and [Yourdon, 1988] for more discussion of these concepts.) The first two of these models are legitimate products of the systems analysis phase of a project; the third should be carried out during the design phase of a project and should be reviewed by people with skills somewhat different from those of a pure systems analyst. Thus, the first question that the reviewers in an analysis walkthrough must ask themselves is: What kind of analysis model are we reviewing?

The true requirements of a system should be described in a form that ignores all technological constraints and imperfections. Address the question: What would the user's requirements be if we had perfect technology— computers that had zero cost, occupied zero physical space, performed calculations in zero time, stored an arbitrarily large amount of data and retrieved any of it in zero time, and that never broke down, malfunctioned, or suffered from dust, dirt, electrical outages, or other vagaries of nature? The reviewers must ask themselves this question over and over again when they review an *essential model* of a system (roughly equivalent to the "new logical model" in such books as [DeMarco, 1978]). It is easy to make assumptions about the use of current hardware/software technology, and those assumptions often show up in subtle ways—not only because the systems analyst is usually surrounded by such technology in his everyday work, but also because the user is becoming increasingly familiar with (and thus brainwashed by) existing computer technology.

Even when the essential model is free of technological assumptions, it still requires thorough, critical, and innovative review by the walkthrough participants. All too often, the essential model will be the user's statement of, "Here's the business policy that I *must* carry out," as compared to the preferable statement of, "Here's a description of the business policy that I really *want* to carry out." Many systems are developed to meet the needs of an individual user or a small department of users, and an objective analysis of their needs often will show that they are carrying out their day-to-day business because they have to, not because they want to. It may be because nobody at a sufficiently high level in the organization has adequately defined what business the entire enterprise is in; or it may be the result of political battles between departments; or it might be the lack of creative problem-solving on the part of the systems analyst.

Thus, the reviewers in a walkthrough of the essential model components should ask themselves, "Are the user requirements merely a means of coping with a problem that they wish they didn't have to deal with in the first place? If so, is there any way of making the problem go away? Is there any way of shifting the problem somewhere else, so that this organization doesn't have to develop a system to cope with it?" This problem does require truly creative thinking, and such books as [De Bono, 1967] and [De Bono, 1976] on lateral thinking might be good background reading for the project team.

Once the essential model has been developed, it will become necessary to review the model derived from it: the *user implementation model*. This model is the essential model together with any implementation constraints which the user feels must be imposed to make the system acceptable. The most obvious constraints involve the choice of an "automation boundary"—i.e., what portions of the essential model will be carried out by humans (both in terms of computations and data storage/retrieval) and what portions by automated facilities? Other constraints typically involve such issues as response time, environmental constraints, and the all-important issues of formats and layouts for inputs and outputs.[1]

The questions raised in the review of a user implementation model are perhaps the most difficult and politically sensitive of all. Essentially, the reviewers must ask if the user-imposed constraint on the implementation of the system is a practicable one. If the user-imposed constraint is the result of government legislation or a contractual obligation imposed by an external organization, then it is probably not negotiable. But if it has been imposed because "that's the way we've always done things," then it should be questioned and challenged—tactfully, politely, and with all the humility appropriate for someone who is not an expert in the user's business area—but challenged nonetheless. If the constraint has been imposed without the user's awareness that it *was* a constraint, then the walkthrough will have done some good!

A.4   Quality

Last is the issue of the overall quality of the requirements model(s). Even if it is complete, internally consistent, and an accurate reflection of the user's requirements, it may still not be an adequate product. In the words of Capers Jones (a nationally prominent expert on issues of software quality and productivity, who developed his ideas at IBM, ITT, and Nolan-Norton before having the good sense to start his own firm), "Quality is never having to say you're sorry." And that is indeed the last question that the

---

[1] The user implementation model is also a way to look at various *alternatives* for implementation and to see what the user thinks of them. The user and analysts can discuss how well the automation boundary has been chosen, what the cost-benefits look like, etc. For this reason, there will often be more than one user implementation model reviewed in a walkthrough. The one with the highest degree of acceptance is the one that will be carried forward to the systems implementation model.

reviewers should ask themselves: *Is there anything about this statement of user requirements that we will someday have to apologize for?*

Since we have already reviewed the user requirements for completeness, consistency, and correctness, we shouldn't have anything substantive to apologize for in the future. Thus, the questions at this stage primarily are concerned with style, organization, legibility, and maintainability of the analysis documents themselves. Thus, here are some appropriate questions:

- *Have the names—process names, names of data elements, names of data stores, names of objects in an entity-relationship diagram, names of states in a state transition diagram—been well chosen?* Are they consistent? Are they cryptic and influenced by the analyst's knowledge of computer technology, or are they business-oriented terms that are meaningful in the user's environment?

- *Is the analysis document well organized?* Does it have an index and/or a table of contents? Has it been packaged in a form that is presentable, such as in an attractive three-ring binder, or is it a pile of loose sheets of handwritten documents?

- *Can the analysis document be maintained?* Is the original document stored in a place known to all concerned? Can it be updated and revised, and are there procedures for doing this in an organized, controlled fashion? (Once again, it becomes evident that automated, CASE-oriented support facilities are the long-range solution for most MIS organizations.)

- *Has the statement of user requirements been factored or decomposed to a sufficiently low level of detail that everyone is confident that the designers or implementers will be able to concentrate on a* solution *to the problem, rather than an ongoing elaboration of the* definition *of the problem?* This decomposition can be accomplished by ensuring that the process specification for each bottom-level bubble of a dataflow diagram is no more than a page or two of material. If a bottom-level bubble of an analysis document says something like "Do payroll" without an accompanying intelligent one-page process specification, then (aside from the fact that it would violate some of the earlier walkthrough guidelines) the designers probably will be coming back to the systems analysts or the end users to ask a lot of questions.

APPENDIX B:

GUIDELINES FOR DESIGN WALKTHROUGHS

This appendix provides specific guidelines for the conduct of systems design walkthroughs—walkthroughs of products that describe the implementation strategy for the system. It assumes that the project team is using structured design techniques of the nature described in [Yourdon and Constantine, 1978] or [Page-Jones, 1988], and that the user requirements have been developed using structured analysis techniques, as described in Appendix A. There are, of course, many other methodologies for structured design, notably the Jackson approach [Jackson, 1975] and the Warnier-Orr approach [Orr, 1977], and the guidelines below may have to be modified slightly for them.

Reviewers in a design walkthrough should look for the same general things that we discussed in Appendix A: *completeness, integrity, correctness,* and *quality.* Each of these is discussed in more detail below.

B.1   Completeness

Just as an analysis walkthrough requires a complete specification, a design walkthrough requires a complete design—either a design of the entire system, or preferably the design of a meaningful subset of the system.

Thus, the reviewers should ask these questions:

- Are the structure charts complete? Are all of the modules named? Are all of the connectors between modules named properly? Are the input/output interfaces between the modules properly named? Is the overall diagram properly labelled or numbered so that you can tell what subset of the overall design it represents?

- Are the design data dictionary definitions complete? Have all of the data elements, flags, and switches been defined? Have the components of those data elements been defined in sufficient detail?

- Does each module in the structure chart have an associated module specification? (Library modules presumably have a pre-written module specification, which the designer may not have included as part of his documentation.)

- Are there proper descriptions of the physical files and/or databases and/or tables with which the modules will interact? These physical file descriptions will be derived from the entity-relationship diagrams in the analysis modules, and they will be described in a form that is appropriate for the kind of database management system, or file management system, or table (or array) that the designer intends to use.

- If the implementation of the system will involve on-line dialogues with users, and/or if the implementation will involve such real-time issues as signals and interrupts, are there appropriate state transition diagrams as part of the design model?

As with the analysis walkthroughs, much of this completeness checking is almost clerical in nature and could be accomplished by an administrative person; alternatively, it could be accomplished by appropriate automated tools.

B.2 Integrity

After ensuring that the design model is complete, the reviewers should examine it carefully to ensure that it does not contain internal contradictions or inconsistencies. Again, much of this review is semi-clerical in nature and could be done with an automated designer's workbench. I assume in this book, though, that the majority of design reviews will be done manually until the mid-1990s.

The major integrity tests that should be performed in a design walkthrough are the following:

- *Undefined data elements*— a structure chart may show a data element (or control flag, etc.) that is passed as a parameter from one module to another; however, an examination of the design data dictionary may show that the data element has not been defined.

- *Undefined files, tables, control blocks, and other data structures*— the structure chart may show a reference from a module to some kind of data structure, a global table, or an external file that is not defined.

- *Undefined references to a data element within a module specification*— the structure chart shows the interfaces between modules, hence it shows the input parameters received by a module, and the output parameters returned to the superordinate "calling" module when the called module exits. These input and output parameters should coincide with the inputs and outputs listed in the module specification (which may have been written in pseudocode, or some other form of program design language, or with a flowchart). If the module specification shows a reference to some other external parameter, as opposed to local variables and constants, then there is an inconsistency. The structure chart and the module specification are in disagreement, and it is important to find out which one is right. This kind of problem is particularly common when the designers have a long history of developing systems in COBOL, FORTRAN, or BASIC, where there is a tendency to assume that *all* parameters will be global and thus accessible to all modules (and thus not shown explicitly on the structure chart).

- *Module specifications whose inputs and outputs do not correspond to the structure chart*—this test is a variation on the problem described above.

- *Balancing errors*—the concept of balancing usually is discussed in textbooks on structured analysis, but it is equally important in the area of design. A module which appears as a bottom-level module in one structure chart may be the top-level module in a lower-level structure chart. For the overall design model to be properly balanced, that module's inputs and outputs must be exactly the same on both diagrams.

- *Errors involving conservation of data*—this problem is analogous to the "conservation of data" problem in the analysis model: A module cannot magically create output if it does not have necessary inputs (we are ignoring, of course, such examples as a "random number generator" module, but even that module usually requires an initial input that acts as a seed for the random number generation process). In any case, the task of the reviewers is to ask whether it is possible for the module to generate its required outputs from the available input parameters.

- *Superfluous input parameters*—again, this problem is analogous to the superfluous data problem in the analysis model; however, it is likely to be a more common problem in design models because the designers are more influenced by their knowledge of technology. Thus, they are more likely to draw a structure chart that shows an entire database record as an input parameter to a module, even if that module only uses one or two fields of data within the record; when questioned, the designer often will mumble vague things about "efficiency" or "that's the way we've always done it."

B.3   Correctness

When the reviewers are convinced that the design model is complete and internally consistent, then they can address the issue of correctness. The question they must ask themselves at this point is quite simple: Is the design a correct implementation of the analysis model?   Will the implementation, as represented by the design model, do what the user wants the system to do as described in the analysis model?

The reviewers must compare the design model— the structure chart, the design data dictionary, the descriptions of physical files and tables, the module specifications, and appropriate state transition diagrams— with the analysis model from which it was derived. They must establish an exact correspondence between each element of the analysis model— each bubble, each data element, and so on— and one or more components of the design model— one or more modules that implement an analysis bubble, a data element in the design data dictionary that was derived from a corresponding data element in the analysis data dictionary, for example. The object is to verify that the design model is an *exact* implementation of the analysis model— nothing more and nothing less.

The "nothing less" problem is an obvious one: If the analysis model has a process bubble for which there is no corresponding module or set of modules in the design structure charts, it means that some *essential* requirement imposed by the user will not be present in the implementation of the system.   Obviously, it is better to discover such a problem during the design review (when the users and project managers may not even be present!) than to have the problem become evident during the final testing of the system.

The "nothing more" problem is a more subtle one and is usually the result of excessive enthusiasm of the designers:

There are one or more modules in the design model for which there are no corresponding process bubbles in the analysis model. The designer has usually said to himself, "Even though they didn't ask for it, I'll bet the users would love to have feature X in their new system." Maybe so; but that is a decision for the user and the systems analyst to make, perhaps with some helpful advice from the designer, but not a decision for the designer to make unilaterally.

Sometimes the problem is more straightforward: The designer has included features in the design in an attempt to implement his understanding of the user's requirements, but the design model and the analysis model do not correspond properly.

B.4   Quality

Finally, the reviewers must consider the overall quality of the design model. Once again, I define quality as "never having to say you're sorry." Most textbooks on structured design, such as [Yourdon and Constantine, 1978], [Page-Jones, 1988], [Jackson, 1975], and [Orr, 1977], have a great deal to say about the quality of a design, and those guidelines can be used to help develop an agenda for this last portion of the design walkthrough. The most important aspects of quality are the following:

* *Cohesion.* Does each module carry out one single, well-defined function, or is it a combination of multiple functions, or some fragment of a function? If the module's purpose can be described in a simple English sentence containing a single, transitive verb, and a single, non-plural object, then the module is probably functionally cohesive. If the designer requires a long, run-on sentence with multiple verbs and objects to describe what the module does, then it is probably not highly cohesive.

* *Coupling.* Is each module reasonably independent of other modules in the system? Can it be modified internally without nasty side-effects elsewhere in

the system? Coupling and cohesion go hand in hand: If the modules in a system are tightly coupled, the cohesion of the modules is generally poor. Since cohesion is often difficult to determine (because it requires the designer to describe, honestly and accurately, what the module does), it is helpful to begin with coupling. If all of the inter-module connections and references have been scrupulously shown on the structure chart, coupling problems usually will stick out like a sore thumb.

- *Span of control.* Do the "manager" modules have too many subordinates? If so, the internal procedural logic of those modules will be extremely complicated— too complicated for the development programmer to code properly or the maintenance programmer to maintain easily. Such problems often can be solved by factoring out some of the management logic into lower level "junior manager" modules. A good rule of thumb, or heuristic, is that a manager module should not have more than 7 ± 2 immediate subordinates, unless the manager serves as a "transaction center," dispatching control to one of N subordinates depending on the value of a transaction code.

- *Scope of control/scope of effect.* In a well-designed system, any module whose behavior is governed by the outcome of a decision will be subordinate to the module that makes the decision. (One can easily imagine an analogous principle within human organizations: A worker whose day-to-day behavior depends on the outcome of a management decision should be somewhere within the empire of the manager who makes the decision.) Violations of this guideline usually lead to a variety of unpleasant design characteristics: excessive flags and switches, duplicated decision-making, or "pathological connections" (e.g., ALTER statements in COBOL) between modules. All of these lead to more difficult testing, debugging, and maintenance.

APPENDIX C:

GUIDELINES FOR CODE WALKTHROUGHS

This appendix provides specific guidelines for the conduct of code walkthroughs. It assumes that the code has been written in a conventional third-generation programming language such as COBOL, PL/I, FORTRAN, C, or Pascal, and that the code has been compiled prior to the walkthrough. Some modifications may be necessary for primitive organizations that still keypunch their source programs on cards or chisel them onto stone tablets prior to compilation.

The issues raised in Appendix A and Appendix B are largely relevant in code walkthroughs, too. The reviewers should be examining the code program for *completeness, integrity, correctness, and quality.*

C.1   Completeness

This check should be a relatively easy job for the reviewers: Either the code is there, or it isn't. However, it's possible that the program being reviewed contains subroutine calls to lower-level modules which exist only as "stubs." In this case, the reviewers will have to decide whether they are willing to review the program before the stubs have been replaced with "real" code.[1]

---

[1] "Stubs" are commonly used in top-down development and testing of programs and systems: top-level modules can be coded and tested before lower-level modules have been coded by substituting a fake lower-level module, known as a stub, which does nothing more than (a) exit immediately, or (b) return a constant output in all cases, or (c) ask for help from the programmer himself at an on-line terminal.

C.2　　Integrity

If the computer program was written in a language like COBOL, Pascal, Ada, C, or Modula-2, we can assume that almost all of the appropriate integrity testing has been performed by the compiler— that is, syntax errors, undefined variables, and other such errors will be flagged by the compiler with error messages. Also, "type" errors will usually be caught— e.g., module A attempting to pass a character string as an input parameter to module B, when module B was expecting a floating point number.

For other programming languages— notably BASIC, FORTRAN, and pre-ANSI versions of C— the integrity testing is not as thorough as it should be. BASIC and FORTRAN, for example, allow the programmer to use a local variable *without* first declaring it and describing its attributes; modern software engineers consider this to be a crime on a par with incest and child abuse.

Indeed, it is doubtful that *any* programming language or *any* compiler will provide complete integrity testing. It would be appropriate, therefore, for the reviewers to first develop a specific list of integrity tests that they will perform manually; or alternatively, the reviewers might decide to develop or purchase a preprocessor that will examine code for such integrity errors *before* it is compiled.

Fagan provides the following list of sample walkthrough review questions for a code walkthrough in [Fagan, 1976]:

1. Is the correct condition tested (If X = ON vs. If X = OFF)?
2. Is (Are) correct variable(s) used for test (If X = ON vs. If Y = ON)?
3. Are null THENs/ELSEs included as appropriate?
4. Is each branch target correct?
5. Is the most frequently exercised test leg the THEN clause?
6. Are absolute (literal) constants used where there should be symbolics?
7. On a comparison of two bytes, should all bits be compared?
8. On built data strings, should they be character or hex?
9. Are internal variables unique or confusing if concatenated?
10. Should any registers be saved on entry? Are they restored on exits?
11. Are all constants defined?
12. Are all increment counts properly initialized (0 or 1)?

C.3   Correctness

The reviewers will spend most of their time in this area. Their job is to review the code to determine whether it is a complete, correct implementation of the design model (the structure chart and associated documents) from which it was derived, *nothing more and nothing less.* Just as designers can become overly enthusiastic and introduce functionality that was not requested in the analysis model, so the programmers can become overly enthusiastic and introduce functions that were not present in the design model. Since code is a tangible implementation of the earlier, abstract models of design and analysis, some of the extra features introduced by the programmer may be rather subtle. The analysis model, for example, may have defined a data element known as Social Security Number as a nine-digit integer; that same definition may have appeared in the design model. But the programmer may have decided to define the Social Security number as a long integer which, because of the hardware features or the language features, can accommodate a twelve-digit number. If questioned about this, the programmer might shrug his shoulders and say, "That's the only way I can make it work," or he might smile craftily and say, "I heard a rumor that the Government is about to run out of available Social Security numbers, and they're going to have to add an extra digit, so I thought I would be prepared," or he might offer to compromise by introducing some error-checking code to ensure that nobody enters a Social Security number longer than 9 digits. Or he might insist that the MIS organization upgrade their facilities to include a proper database management system (so that the Social Security number could be defined *once* for all programs that use it, with error-checking built in), and a proper programming language.

C.4   Quality

At this point, any further discussion of the program by the reviewers typically will involve issues of style. The difference between good style and bad style depends on the programming language and on traditions and standards within the organization; in some cases, it may also be affected by contractual obligations—for example, if the organization is writing programs to be delivered to an external customer such as the U.S. Defense Department.

A variety of books, such as [Yourdon, 1975] and [Kernighan and Plauger, 1974], discuss programming style in great detail; typically the reviewers will want to examine the following aspects of the program for proper style:

- *Data names*—are they cryptic names like X7, or are they "long" names like CUSTOMER_ADDRESS? Do the names provide a reasonable description of the meaning and content of the data?

- *Use of language features*—some languages, like COBOL, PL/I, and Ada, are so rich in features that an MIS organization may want its programmers to use a subset of the language. After all, if the language provides nineteen different ways to add two integers, it makes sense to standardize on one or two of those ways.

- *Formatting*—the code should be neatly organized. Levels of loops such as DO-WHILE statements, and IF-THEN-ELSE statements should be indented properly; modules such as subroutines, procedures, and COBOL SECTIONs should begin at the top of a page, and so forth.

- *Complexity*—the program should be simple enough so that the average programmer can read it and understand it. There are various methods for determining program complexity (see, for example, [DeMarco, 1982] and [Halstead, 1977]); some organizations may be content with such simple approaches as restricting the length of a program module to one page of a listing, or forbidding more than three levels of nested IF statements.

- *Appropriate comments and documentation*—as Kernighan and Plauger said, "Document unto others as you would have them document unto you"

[Kernighan and Plauger, 1974]. Too little documentation is an obvious problem. Too much documentation, especially redundant, trivial documentation, also can be a problem.

* *Use of structured programming constructs*— it is inconceivable that anyone is still writing "rat's nest" unstructured code in these modern times; after all, the 20th century is almost over! However, the *method* of writing structured code in such primitive languages as FORTRAN and COBOL may require some agreement on proper style. For example, how should an IF-THEN-ELSE construct be coded in a language that doesn't permit an ELSE clause? There are a dozen different ways of doing it; the programmers should agree on one or two acceptable ways, and then ensure in the walkthrough that the style has been followed properly.

* *Issues of style*— A variety of commercial programs, such as PMETRIC and CMETRIC, can be used to review those aspects of a program that will make the code difficult to maintain and understand. Such style analyzers typically "score" a program based on such things as the average variable name length, the use of indentation (which can be accomplished automatically with "pretty print" programs), the use of GOTOs, and the mix of comments. For more discussion of style analysis, see [Rees, 1982] and [Jorgensen, 1980].

BIBLIOGRAPHY

[Aron, 1983]   *The Program Development Process, Part II: The Programming Team.* Joel Aron (Reading, MA: Addison-Wesley, 1983).

[Aukee, Beckwith, and Buttenmiller, 1973]   *Inside the Management Team.* Waino E. Aukee, John A. Beckwith, and Karl O. Buttenmiller, eds. (Kansas City, KS: Interprint Publishers, 1973).

[Bass, 1975]   *Management by Task Forces: A Manual on the Operation of Interdisciplinary Teams.* Lawrence W. Bass (New York: Lomond, 1975).

[Berber and Oktaba, 1987]   "Crafting Reusable Software in Modula-2." René Berber and Hanna Oktaba (*Byte*, September 1987).

[Berne, 1961]   *Transactional Analysis in Psychotherapy.* Eric Berne (New York: Grove Press, 1961).

[Berne, 1963]   *The Structure and Dynamics of Organizations and Groups.* Eric Berne (New York: J.B. Lippincott, 1963).

[Berne, 1964]   *Games People Play.* Eric Berne (New York: Ballantine Books, 1964).

[Boar, 1984]   *Application Prototyping.* Bernard Boar (New York: John Wiley & Sons, 1984).

[Boehm, 1980]   "Developing Small-Scale Application
                     Software Products: Some Experimental
                     Results." Barry Boehm (*Proceedings, IFIP
                     8th World Computer Congress*, October
                     1980, pp. 321-326).

[Boehm, 1981]   *Software Engineering Economics*. Barry
                     Boehm (Englewood Cliffs, NJ: Prentice
                     Hall, 1981).

[Brill, 1983]   *Building Controls into Structured
                     Systems*. Alan E. Brill (Englewood Cliffs,
                     NJ: Yourdon Press/Prentice Hall, 1983).

[Brod, 1984]   *Techno Stress*. Craig Brod (Reading, MA:
                     Addison-Wesley, 1984).

[Brooks, 1975]   *The Mythical Man-Month*, Fred Brooks
                     (Reading, MA: Addison-Wesley, 1975).

[Connell and   *Rapid Prototyping*. John Connell and
Shafer, 1988]   Linda Shafer (Englewood Cliffs, NJ:
                     Yourdon Press/Prentice Hall, 1988).

[Cougar and   *Motivating and Managing Computer
Zawacki, 1980]   Personnel*, J. Daniel Cougar and Robert A.
                     Zawacki (New York: John Wiley & Sons,
                     1980).

[Crossman, 1979]   "Some Experiences in the Use of
                     Inspection Teams in Applications
                     Development." T. D. Crossman (SHARE,
                     Inc. *Proceedings, Application Devel-
                     opment Symposium*, October 1979, pp.
                     163-168).

[Curtis, Sheppard,   "Third Time Charm: Stronger Prediction
and Milliman,   of Programmer Performance by
1979]   Software Complexity Metrics." B. Curtis,
                     S. Sheppard, and P. Milliman (*IEEE

Proceedings of the Fourth International Conference on Software Engineering, 1979, pp. 356-360).

[De Bono, 1967]   *The 5 Day Course in Thinking.* Edward De Bono (London: Penguin Books, 1967).

[De Bono, 1976]   *Teaching Thinking.* Edward De Bono (London: Penguin Books, 1976).

[DeMarco, 1978]   *Structured Analysis and System Specification.* Tom DeMarco (Englewood Cliffs, NJ: Yourdon Press/Prentice Hall, 1978).

[DeMarco, 1979]   *Concise Notes on Software Engineering.* Tom DeMarco (Englewood Cliffs, NJ: Yourdon Press/Prentice Hall, 1979).

[DeMarco, 1982]   *Controlling Software Projects.* Tom DeMarco (Englewood Cliffs, NJ: Yourdon Press/Prentice Hall, 1982).

[DeMarco and Lister, 1987]   *Peopleware: Productive Projects and Teams.* Tom DeMarco and Timothy Lister (New York: Dorset House, 1987).

[Drury, 1984]   *Assertive Supervision: Building Involved Teamwork.* Susanne S. Drury (Los Angeles, CA: Res Press, 1984).

[Dunn, and Ullman, 1982]   *Quality Assurance for Computer Software.* Robert Dunn and Richard Ullman (New York: McGraw-Hill, 1982).

[Dunn, 1984]   *Software Defect Removal.* Robert Dunn (New York: McGraw-Hill, 1984).

[Dyer, 1977]   *Team Building: Issues and Alternatives.* William G. Dyer (Reading, MA: Addison-Wesley, 1977).

[Ends and
Page, 1984]

Organizational Team Building. Earl J.
Ends and Curtis W. Page (Lanham, MD:
University Press of America, 1984).

[Fagan, 1976]

"Design and Code Inspections to Reduce
Errors in Program Development." Michael
Fagan (*IBM Systems Journal*, Vol. 15, No.
3, July 1976, pp. 182-211; also reprinted
in [Yourdon, 1982]).

[Fagan, 1986]

"Advances in Software Inspections."
Michael Fagan (*IEEE Transactions on
Software Engineering*, Vol. SE-12, No. 7,
July 1986, pp. 744-751).

[Flavin, 1981]

*Fundamental Concepts of Information
Modeling.* Matt Flavin (Englewood Cliffs,
NJ: Yourdon Press/Prentice Hall, 1981).

[Ford, 1983]

*The ABCs of Managing with Employee
Teams.* Sondra Ford (S. Ford and
Associates, 1983).

[Francis, and
Young, 1979]

*Improving Work Groups: A Practical
Manual for Team Building.* Dave Francis
and Don Young (San Diego, CA: University
Associates, 1979).

[Gane and
Sarson, 1977]

Structured Systems Analysis. Chris Gane
and Trish Sarson (New York: Improved
Systems Technologies, 1977).

[Gargaro, 1987]

"Reusability Issues in Ada." Anthony
Gargaro (*IEEE Software*, July 1987).

[Goodman, 1987]

The Complete HyperCard Handbook.
Danny Goodman (New York: Bantam
Books, 1987).

[Grady and
Caswell, 1987]

*Software Metrics: Establishing a Company-
Wide Program.* Robert B. Grady and

Deborah L. Caswell (Englewood Cliffs, NJ: Prentice Hall, 1987).

[Halstead, 1977]   *Elements of Software Science.* Maurice Halstead (New York: Elsevier Press, 1977).

[Hansen, 1984]   *Data Structured Program Design.* Kirk Hansen (Topeka, KS: Orr & Associates, 1984).

[Hausen, 1984]   *Software Validation: Inspection-Testing-Verification-Alternatives.* H. L. Hausen (New York: Elsevier Press, 1984).

[Higgins, 1979]   *Program Design and Construction.* David Higgins (New York: Prentice-Hall, 1979).

[Howe, 1983]   *Data Analysis for Data Base Design.* D. R. Howe (Baltimore, MD: Edward Arnold, 1983).

[Inmon, 1988]   *Information Engineering for the Practitioner.* William Inmon (Englewood Cliffs, NJ: Yourdon Press/Prentice Hall, 1988).

[Jackson, 1975]   *Principles of Program Design.* Michael Jackson (New York: Academic Press, 1975).

[Jackson, 1983]   *System Development.* Michael Jackson (Englewood Cliffs, NJ: Prentice Hall, 1983).

[Jones, 1977]   "Program Quality and Programmer Productivity." T. Capers Jones (IBM Technical Report TR 02.764, January 28, 1977).

[Jones, 1986]   *Programming Productivity.* T. Capers Jones (New York: McGraw-Hill, 1986).

[Jorgensen, 1980] "A Methodology for Measuring the Readability and Modifiability of Computer Programs," A. H. Jorgensen (*BIT*, Volume 20, 1980, pp. 394-405).

[Keller, 1987] *Expert System Technology: Development and Application.* Robert Keller (Englewood Cliffs, NJ: Yourdon Press/Prentice Hall, 1987).

[Kernighan and Plauger, 1974] *The Elements of Programming Style.* Brian W. Kernighan and P. J. Plauger (New York: McGraw-Hill, 1974).

[Lewis, 1978] "Group, Not Lines, Builds Sewing Machines." Paul Lewis (*New York Times*, March 18, 1978).

[Lientz and Swanson, 1980] *Software Maintenance Management.* B. P. Lientz and E. B. Swanson (Reading, MA: Addison-Wesley, 1980).

[Martin, 1984] *An Information Systems Manifesto.* James Martin (Englewood Cliffs, NJ: Prentice Hall, 1984).

[Martin and McClure, 1985] *Structured Techniques for Computing.* James Martin and Carma McClure (Englewood Cliffs, NJ: Prentice Hall, 1985).

[McClellan, 1984] *The Coming Computer Industry Shakeout.* Stephen McClellan (New York: John Wiley & Sons, 1984).

[McCulloch, 1982] *Performance Teams: Completing the Feedback Loop.* Janis McCulloch (Performance Management, 1982).

[McMenamin and Palmer, 1984] *Essential Systems Analysis.* Steve McMenamin and John Palmer (Englewood Cliffs, NJ: Yourdon Press/Prentice Hall, 1984).

[Mills, Linger, and Hevner, 1986] *Principles of Information Systems Analysis and Design,* Harlan D. Mills, Richard Linger, and Alan B. Hevner (New York: Academic Press, 1986).

[Myers, 1976] *Software Reliability.* Glenford Myers (New York: John Wiley & Sons, 1976).

[Myers, 1978] "A Controlled Experiment in Program Testing and Code Walkthroughs-Inspections." Glenford Myers (*Communications of the ACM,* September 1978, pp. 760-768).

[Myers, 1979] *The Art of Software Testing.* Glenford Myers (New York: John Wiley & Sons, 1979).

[Naisbitt, 1983] *Megatrends.* John Naisbitt (New York: Warner Books, 1983).

[Naisbitt and Aburdene, 1985] *Reinventing the Corporation: Transforming Your Job and Your Company for the New Information Age.* John Naisbitt and Patricia Aburdene (New York: Warner Books, 1985).

[Neumann, 1985] "Some Software Disasters and Other Egregious Horrors." Neumann (*ACM Software Engineering Notes,* January 1985).

[Orr, 1977] *Structured Systems Development.* Ken Orr (Englewood Cliffs, NJ: Yourdon Press/Prentice Hall, 1977).

[Orr, 1981]   *Structured Requirements Definition*. Ken
                       Orr (Topeka, KS: Orr & Associates, 1981).

[Ouchi, 1982]   *Theory Z Management*. William Ouchi
                       (New York: Avon Books, 1982).

[Page-Jones,   *The Practical Guide to Structured Systems*
1988]   *Design*, 2nd ed. Meilir Page-Jones
                       (Englewood Cliffs, NJ: Yourdon
                       Press/Prentice Hall, 1988).

[Peters, 1987]   *Thriving on Chaos*. Tom Peters (New
                       York: Alfred Knopf, 1987).

[Rees, 1982]   "Automatic Assessment Aids for Pascal
                       Programs." M.J. Rees (*ACM SIGPLAN
                       Notices*, October 1982, pp. 33-42).

[Rimler and   *Small Business: Developing the Winning*
Humphreys,   *Management Team*. George W. Rimler
1980]   and Neil J. Humphreys (UMI, 1980).

[Roeske, 1983]   *The Data Factory*. Edward Roeske
                       (Englewood Cliffs, NJ: Yourdon
                       Press/Prentice Hall, 1983).

[Sackman,   "Exploratory Experimental Studies
Erickson, and   Comparing Online and Offline
Grant, 1968]   Programming Performance." H. Sackman,
                       W. J. Erickson, and E. E. Grant
                       (*Communications of the ACM*, January
                       1968, pp. 3-11).

[Scott, 1987]   "Can Programmers Reuse Software?" Del
                       T. Scott (*IEEE Software*, July 1987).

[Sculley, 1987]   *Odyssey*. John Sculley (New York: Harper
                       & Row, 1987).

[Semprevivo, 1981]   *Teams in Information Systems Development.* Philip Semprevivo (Englewood Cliffs, NJ: Yourdon Press/ Prentice Hall, 1981).

[Thayer, Lipow, and Nelson, 1978]   *Software Reliability: A Study of Large Project Reality.* T. A. Thayer, M. Lipow and E. C. Nelson (New York: North-Holland, 1978).

[Thomsett, 1981]   *People and Project Management.* Rob Thomsett (Englewood Cliffs, NJ: Yourdon Press/Prentice Hall, 1981).

[Townsend, 1984]   *Further up the Organization,* Robert Townsend (New York: Alfred Knopf, 1984).

[Tracz, 1987]   "Reusability Comes of Age," Will Tracz (*IEEE Software,* July 1987).

[Turkle, 1984]   *The Second Self.* Sherry Turkle (New York: Simon & Schuster, 1984).

[Tuttle and Sink, 1985]   "Taking the Threat Out of Productivity Measurement." T. Tuttle and D. Sink (*National Productivity Review,* Winter 1984-85, pp. 24-32).

[Ward, 1984]   *Systems Development Without Pain.* Paul Ward (Englewood Cliffs, NJ: Yourdon Press/Prentice Hall, 1984).

[Ward and Mellor, 1985]   *Structured Development for Real-Time Systems.* Paul Ward and Steve Mellor (Englewood Cliffs, NJ: Yourdon Press/Prentice Hall, 1985).

[Weinberg, 1971]   *The Psychology of Computer Programming,* Gerald Weinberg (New York: Van Nostrand, Reinhold, 1971).

[Weinberg and Freedman, 1982] *Handbook of Walkthroughs and Inspections.* Gerald Weinberg and Daniel Freedman (Boston, MA: Little Brown, 1982).

[Yourdon, 1975] *Techniques of Program Structure and Design.* Edward Yourdon (Englewood Cliffs, NJ: Prentice Hall, 1975).

[Yourdon and Constantine, 1978] *Structured Design.* Edward Yourdon and Larry Constantine (Englewood Cliffs, NJ: Yourdon Press/ Prentice Hall, 1978).

[Yourdon, 1979] *Classics in Software Engineering.* Edward Yourdon, ed. (Englewood Cliffs, NJ: Yourdon Press/Prentice Hall, 1979).

[Yourdon, 1982] *Writings of the Revolution: Selected Readings in Software Engineering.* Edward Yourdon, ed. (Englewood Cliffs, NJ: Yourdon Press/Prentice Hall, 1982).

[Yourdon, 1985] *Managing the Structured Techniques,* 3rd ed. Edward Yourdon (Englewood Cliffs, NJ: Yourdon Press/Prentice Hall, 1985).

[Yourdon, 1986] *Nations at Risk.* Edward Yourdon (Englewood Cliffs, NJ: Yourdon Press/Prentice Hall, 1986).

[Yourdon, 1988] *Managing the System Life Cycle,* 2nd ed. Edward Yourdon (Englewood Cliffs, NJ: Yourdon Press/Prentice Hall, 1988).

[Yourdon, 1989] *Modern Structured Analysis.* Edward Yourdon (Englewood Cliffs, NJ: Yourdon Press/Prentice Hall, 1989).

INDEX